ENG SDT 30
KU-429-042

New Methods for Modelling Processes within Solids and at their Surfaces

New Methods for Modelling Processes within Solids and at their Surfaces

Edited by

C. R. A. Catlow
Royal Institution, London

A. M. Stoneham
AEA Technology, Harwell

and

Sir John Meurig Thomas
Royal Institution, London

TRINITY SCIENCE LIBRARY

THE ROYAL SOCIETY
OXFORD UNIVERSITY PRESS
1993

Oxford University Press, Walton Street, Oxford OX2 6DP
Oxford New York Toronto
Delhi Bombay Calcutta Madras Karachi
Kuala Lumpur Singapore Hong Kong Tokyo
Nairobi Dar es Salaam Cape Town
Melbourne Auckland Madrid
and associated companies in
Berlin Ibadan

Oxford is a trade mark of Oxford University Press

Published in the United States
by Oxford University Press Inc., New York

© *The Royal Society 1993*
First published in Philosophical Transactions of the Royal Society, *vol 341, pp. 193–371* (*1992*)

All rights reserved. No part of this publication may be reproduced, stored in a retrieval system, or transmitted, in any form or by any means, without the prior permission in writing of Oxford University Press. Within the UK, exceptions are allowed in respect of any fair dealing for the purpose of research or private study, or criticism or review, as permitted under the Copyright, Designs and Patents Act, 1988, or in the case of reprographic reproduction in accordance with the terms of licences issued by the Copyright Licensing Agency. Enquiries concerning reproduction outside those terms and in other countries should be sent to the Rights Department, Oxford University Press, at the address above.

This book is sold subject to the condition that it shall not, by way of trade or otherwise, be lent, re-sold, hired out, or otherwise circulated without the publisher's prior consent in any form of binding or cover other than that in which it is published and without a similar condition including this condition being imposed on the subsequent purchaser.

A catalogue record for this book is available from the British Library

Library of Congress Cataloging in Publication Data

New methods of modelling processes within solids and at their surfaces
/edited by C.R.A. Catlow, A.M. Stoneham, and John Meurig Thomas.
1. Solids—Mathematical models. 2. Solids—Computer simulation. 3. Surfaces (Physics)—Mathematical models. 4. Surfaces (Physics)—Computer simulation. 5. Semiconductors—Defects. I. Catlow, C.R.A. (Charles Richard Arthur), 1947–. II. Stoneham, A.M. III. Thomas, J.M. (John Meurig)
QC176.N49 1993 620.1′1—dc20 92–42240

ISBN 0–19–853988–6

Typeset by the University Press, Cambridge
Printed in Great Britain at the
University Press, Cambridge

Preface

More than half a century has passed since the Hartrees' pioneering self-consistent calculations of electronic structure. It is also over half a century since Mott and Littleton first showed how the polarization and deformation due to charged defects could be handled self-consistently. Even the simultaneous optimization of electronic structure and atomic displacements has been achieved in simpler cases for over two decades. Self-consistent studies of excited states, of surface defects, and even of some quite complex atomistic process have been done for more than a decade. So what is new?

First, there are new approaches which give both real advantages and the confidence that previously difficult problems can be managed. The use of Local Density Functional methods, and the Car–Parrinello method are good examples. Their impact has been both through their own strengths and because they have stimulated a whole range of alternatives, all with their strengths. Secondly, the implications of theory have been pursued with vigour. It is not enough to predict one-electron energies, nor even to provide energy levels and wavefunctions alone. The proper role of good theory is to take such calculations further, so that they would predict what had been observed, or could be observed. Thirdly, computer power has developed dramatically: the calculations which demanded the largest computers twenty years ago can be done now on an unsophisticated personal computer. The calculations which can be done on the largest computers today are more far-reaching, and are one of the reasons for this Meeting. But it is not just computer power which has changed: computer displays have altered in style and concept. Colour is not just decoration, but a means to interpret and understand; computer graphics make decisions about future directions visible in the mass of output. Finally, there is a new enthusiasm to give computer modelling impact. This we hope to encourage. It is our belief that, just as modelling clarifies and extends applied science, so do applications enrich the more basic studies by raising questions which would otherwise remain unasked.

Modelling means different things to different scientists. Approaches range from the most complex methods for small numbers of electrons to approaches based on continuum electromagnetics or elasticity theory. To some the words *a priori* are a virtue (like many virtues, more often claimed than demonstrated); to others a degree of empiricism is acceptable. Both classes are represented at this meeting, for they both have their place: today's empirical calculations may be tomorrow's arena for methods which are closer to first principles. The empirical approaches show what can be done; they are a guide for experimental science and a framework for application. We also recognize that studies at an atomic scale alone and their macroscopic consequences are often hard to connect. Indeed, the proper analysis at a mesoscopic level may be an essential link, and this too forms part of the meeting. Such a link is especially important where the cross-fertilization of science and technology is exploited. It is our hope that this Discussion Meeting will help those who apply science to realize what is now possible, and those in basic science to appreciate the diversity of challenge and opportunity the new technologies are providing.

London

October 1992

C. R. A. C.
A. M. S.
J. M. T.

Contents

Understanding defects in semiconductors: spin-off from technology

By G. A. Baraff

AT&T Bell Laboratories, Murray Hill, New Jersey 07974, U.S.A.

Defects in semiconductors play a controlling role in determining their technologically useful properties. Particularly important in this regard are defects that introduce levels far from the conduction and valence band edges. Capture or emission of electrons or holes into such levels generally results in displacement of the atoms in the immediate vicinity of the defect, and the action of the defect in determining the semiconductor properties must be understood with this complication in mind. Examples of the role of calculations in establishing some of our currently useful knowledge are presented.

1. Introduction

This paper provides an introduction to the work on deep level defects that constitutes approximately one quarter of this volume. Although the volume is about processes, there is a connection between deep level defects and processes. That connection comes about via phenomena that occur at defects and because of the presence of defects. Defects are dynamic entities that can migrate through the crystal. They can combine with each other in ways similar to the way atoms combine in ordinary chemical processes in gases. Under the influence of external stimuli such as light, or pressure, or exchange of electrons with the environment, they can alter the arrangement of atoms in their immediate vicinity. These changes give rise to some interesting phenomena, namely, large lattice relaxation, negative U, metastability and persistent photoconductivity (Pantelides 1986).

There is an important link between technologically important issues and the study of point defects. Some of this will be mentioned in §2. In §3, I review some basic ideas that underlie what is done in trying to calculate the properties of defects. This is in the nature of an elementary tutorial but it contains notions which, although understood by practitioners in the field, have not been spelled out for a general audience. Section 4 will briefly mention how progress in the ability to calculate defect properties has had an effect on the way we think of them.

2. Technologically driven problems

(a) *Diffusion*

Semiconductor devices are made by successively depositing layers of materials of varying composition and doping on a substrate, sometimes etching away parts of various layers or masking off the areas where new material is to be deposited. During processing, dopants will move around if they are mobile at processing temperatures. The dopant atoms themselves are defects in the lattice. Their diffusion may well be

Phil. Trans. R. Soc. Lond. A (1992) **341**, 195–202
© 1992 The Royal Society

mediated by other defects; vacancies, for example, with which they can trade places, or native interstitials, which eject the dopants from their lattice sites. The energetics involved in these defect reactions control the diffusion. There is therefore interest in being able to calculate the energetics so as to improve diffusion modelling, essential in the design of better manufacturing processes.

(b) *Persistent photoconductivity and the DX centre*

Aluminium gallium arsenide alloy, or AlGaAs, is an important semiconductor material in devices where high speed, low noise, low power consumption and generation or detection of light may be important. In AlGaAs, the DX centre is the lowest energy state of the donor atom when the aluminium fraction exceeds 22% (Mooney 1991), and thus this centre determines the conductivity of the material. Yet, at low temperatures, one cannot count on knowing the conductivity of the material: the DX level has large barriers to both electron capture and emission. At low temperatures, these barriers are large enough that the free electron concentration does not come into equilibrium with the donors and the conductivity of the material then depends on its thermal history. Even worse, persistent photoconductivity is observed at low temperature. When the sample is exposed to light at low temperature, the DX centres are ionized. However, when the light is turned off, the electrons remain in the conduction band for minutes, hours, or days, depending on the temperature and alloy concentration. This is clearly a disaster, both for a photodetector, and for other devices such as the MODFET (a modulation doped field effect transistor). Both of these aspects, the high capture barriers and the persistent photoconductivity, are known to result from the fact that the defect changes its geometric configuration, that is, it undergoes a large lattice relaxation, when its charge state changes. Exactly what the defect is and how it causes these peculiar effects were major intellectual mysteries and major technological annoyances until the theoretical work carried out by Chadi & Chang (1989).

In the case of the MODFET, the operating characteristic (the current versus the potential difference) is determined in large part by the amount of band bending. This is affected in turn by the spatial distribution of ionized donors. At low temperatures, these donors cannot easily recapture their electrons, and thus the operating characteristics of the MODFET changes because the donors are DX centres.

(c) *Production of semi-insulating gallium arsenide and EL2*

When devices are fabricated, material is deposited on a substrate. That substrate has to be semi-insulating to insulate properly the various parts of a device from each other. Equally important, where many transistors are on a single monolithic substrate, a semi-insulating substrate is needed to insulate the devices from each other.

Gallium arsenide, as grown, often contains shallow acceptor levels. Their energy is only slightly above the top of the valence band, and so they are easily occupied by electrons from that band, leaving behind holes that give the material p-type conductivity. However, if mid-gap donors are introduced in greater concentration than the acceptors, they compensate the acceptor levels, filling them all with electrons. Under these conditions, there can be no electrons in the conduction band (they would be captured by the empty mid-gap levels) nor holes in the valence band (they would be filled by electrons from the occupied mid-gap levels). The material becomes semi-insulating.

GaAs has been doped with chromium to make it semi-insulating. However, at the temperatures used in device fabrication, chromium is mobile. This is unacceptable when it is necessary to make some regions semi-insulating, and other regions n-type or p-type. Thus, chromium is unsuited for this purpose. But other mid-gap donors are known. EL2 is one such midgap donor (Martin & Makram-Ebeid 1986). It is immobile at device processing temperatures. It does happen to be metastable at low temperatures. Unlike DX, its metastability is only optically driven. It is suited for use in producing semi-insulating GaAs. Aside from its metastability, which is a fascinating topic, there was technological need to understand the microscopic structure of EL2. The need arose because technologists, although confident that they could make wafers with reproducible EL2 concentration, feared that if there were variations in the trace compositions of the source materials used in production, their recipes for producing wafers with uniform EL2 concentration might fail. Knowing the microscopic identity of EL2 would provide confidence that the recipes could be changed if the source material changed.

(*d*) *Degradation*

Since the properties of devices are controlled by the positions of the dopant atoms, it is necessary that they stay where they are first put, not only during manufacture, but also while the device is in use. The presence of free carriers can cause enhanced mobilities and can cause defect reactions to occur, even at temperatures where these processes would not normally be important. Electrons in the conduction band, for example, liberate energy when they are captured by a deep-lying defect level, and this energy can substitute for the energy that would otherwise have to be supplied thermally. Thus, a device that has a long life on the shelf can fail in use. One needs to understand these enhanced processes as well. This understanding can be aided by precise calculations of the energetics of defect reactions.

3. Basic ideas in deep level studies

(*a*) *Tools and approximations*

The generic problem is to describe the quantum mechanical state of a defect in an otherwise perfect crystal. The defect is an atom or several atoms that are wrong for the crystal, either in their chemical identity, or their location, or both. The relevant variables are the positions of all the electrons in the system and all the atomic cores in the system, but almost universally the Born–Oppenheimer approximation is brought in at this stage. Density-functional theory is most often used to describe how the electrons interact with each other, and norm-conserving pseudopotentials are used to describe how the electrons interact with the atomic cores. Finally, the region of the crystal near the effect is singled out for special attention. One way is by using Green's function. This is a technique that allows one to do a complete calculation even though the region in which the wave functions must be computed is limited to the region around the defect where the potential differs from that of the perfect crystal (Koster & Slater 1954; Baraff & Schluter 1979; Bernholc *et al.* 1980; Kelly & Car 1992).

Another way that has recently been resurrected is the large unit-cell method, where a periodic array of defects is considered. Each defect in the array is surrounded by a unit cell of perfect crystal, and these unit cells are chosen to be large enough that a defect in one unit cell is not influenced by the defect in the next cell. The shift in preferred calculational method from large unit cell to Green's function back to large unit cell (either alone or used with a Green's function) was driven both by the changes in hardware and software: recent advances in calculational techniques for the large unit cell problem, such as Car–Parrinello and conjugate gradient techniques (Car & Parrinello 1985; Stitch *et al*. 1989) were used in defect calculations as well. The results of such calculations are wave functions for the various electronic defect states, and a total energy. This total energy, in the Born–Oppenheimer approximation, depends on the chemical potential μ, on the positions of the atomic cores R_i, on the number of electrons N occupying the defect levels, and on j, the state of excitation of the defect.

(*b*) *Levels*

There are many uses of the word 'level' in quantum mechanics. An important use is in the discussion of defects, where a level is the difference in ground state energies between two systems that are identical, except that the second contains one more electron than the first. This use is important because it defines the conditions under which a defect will be found in one or another equilibrium charge state. To see how this comes about consider the probability of finding a system with a given number of particles as given by the rules of equilibrium statistical mechanics. The system we are considering is the defect, and the occupancy of any or all of the localized states associated with it. The probability P that at temperature T and chemical potential μ the system will be found to contain N electrons is given by

$$P(N,T) = Z^{-1} g_N \exp[-(E_N - \mu N)/kT], \tag{1}$$

where

$$Z \equiv \sum_{N=0} g_N \exp[-(E_N - \mu N)/kT]$$

is the partition function. The ground state energy for the system containing N particles is E_N and the degeneracy factor is g_N. The ratio of probabilities that the system has two electrons rather than one is

$$\begin{aligned}\frac{P(2,T)}{P(1,T)} &= \frac{g_2 \exp[-(E_2 - 2\mu)/kT]}{g_1 \exp[-(E_1 - \mu)/kT]} \\ &= (g_2/g_1)\exp\{-[(E_2 - E_1) - \mu]/kT\} \\ &= (g_2/g_1)\exp[-(\epsilon_2 - \mu)/kT],\end{aligned} \tag{2}$$

where

$$\epsilon_2 \equiv E_2 - E_1. \tag{3}$$

(We assume here that excited states are located at appreciably higher energies. If they are not, it is trivial to include them in the argument being made.) The ratio of probabilities is thus given by an expression which involves a comparison between μ, the chemical potential, and the level ϵ_2, defined as the difference between the two ground state energies. If $\mu < \epsilon_2$, this ratio is nearly zero at low temperature and the defect will be found to have zero or one electron. Conversely, if $\mu > \epsilon_2$, the ratio will be nearly infinite and the effect will be found to be occupied by at least two electrons. This idea is illustrated in figure 1*a*, which shows the levels associated with an isolated vacancy in silicon, at least as they were understood before 1979 (Watkins *et al*. 1979).

The levels shown as lying between the conduction band and the valence band denote those values of μ for which the occupancy of the defect changes. The regions

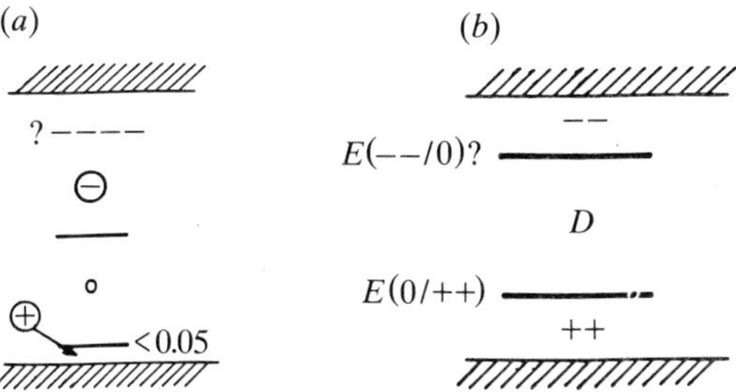

Figure 1. (*a*) Energy levels for the silicon vacancy as they were understood before 1979. (*b*) Energy levels for the silicon vacancy after 1979.

between the levels are labelled by the equilibrium charge state of the vacancy when the chemical potential is anywhere in the region.

(*c*) *Two electron levels: negative U*

Since 1980, it has become clear that the actual situation for the silicon vacancy is as shown in figure 1*b*, namely, if μ is below the lowest level, there are no electrons on the defect and its charge is $+2$, while if μ is above the lowest level, then there are two electrons on the defect and it is neutral (Newton *et al.* 1983). In this case, μ determines an equilibrium occupancy of either zero or two electrons, and this is said to be a 'two electron level'. There is no value of μ for which the system (in equilibrium with a reservoir of charge) will be found with one electron.

To understand how this situation is described by statistical mechanics, consider a system that can hold zero, one or two electrons. The levels are still defined as the differences in ground state energies:

$$\epsilon_1 \equiv E_1 - E_0, \quad \epsilon_2 \equiv E_2 - E_1. \tag{4a, b}$$

It usually takes more energy to add the second electron than the first because the second one is repelled by the one already there. So if we define

$$\epsilon_2 = \epsilon_1 + U \tag{5}$$

then U, the additional energy needed to add the second electron, is expected to be of the order of $e^2/\epsilon R$, where R is the spatial extent of the wave function for the added electron and ϵ is the dielectric constant of the crystal. The partition function in this case takes the form

$$\begin{aligned} Z &= g_0 \exp(-E_0/kT) + g_1 \exp[-(E_1-\mu)/kt] + g_2 \exp[-(E_2-2\mu)/kt] \\ &= g_0 \exp(-E_0/kT)[1+(g_1/g_0)\exp[-(\epsilon_1-\mu)/kT] + (g_2/g_0)\exp\{-2[\bar{\epsilon}-\mu]/kT\}], \end{aligned} \tag{6}$$

where

$$\bar{\epsilon} \equiv \tfrac{1}{2}(\epsilon_1+\epsilon_2). \tag{7}$$

Two energies determine which of the three terms is large: ϵ_1 and $\bar{\epsilon}$. Occupancy will be decided by the placement of μ relative to ϵ_1 and $\bar{\epsilon}$. Although it most often happens that $\epsilon_2 > \epsilon_1$, if the situation is reversed, then by definition of (5), it follows that $U < 0$. This indicates that something other than electrostatic repulsion is active in determining the relative energy needed to add successive electrons. It will also follow, from (4), that

$$2E_1 > E_0 + E_2. \tag{8)}$$

This inequality is what Chadi & Chang (1989) calculated to be the case for the DX defect. In the case of DX, the neutral defect is the one with a single electron, while zero or two electrons correspond to a charge of plus or minus one respectively. In the case of DX, even without a reservoir of electrons to equilibrate with, a collection of neutral defects that are able to exchange electrons among themselves will have a lower energy if half of them have two electrons and the other half have zero electrons. Experiments to establish whether or not this is true for DX have occupied great attention recently and there is now general consensus that this is true.

Another consequence follows from the ordering $\epsilon_2 < \epsilon_1$. In this case, from (7), we have $\bar{\epsilon} < \epsilon_1$. Then, at low temperatures, $P(0,T) \approx 1$ for $\mu < \bar{\epsilon}$, and $P(2,T) \approx 1$ for $\mu > \bar{\epsilon}$, and $P(1,T) \approx 0$ for all μ. The system will, depending on the relative positions of μ and $\bar{\epsilon}$, contain either zero or two electrons if it is in contact with an external reservoir. This describes the two-electron level. Occupancy of the defect changes by two electrons as μ is raised. The silicon vacancy, with its two electron level, is sometimes referred to as being a 'negative U' system, because the two concepts, negative U and two-electron level, are identical.

(d) *Large lattice relaxation and negative U*

Lattice relaxation is the cause of negative U. A connection between the two can be illustrated with a very simple model. Suppose that there is only one active lattice coordinate, say Q. Suppose further that two electrons can go into the same spatial electronic wave function, one with spin up and the other with spin down. Assume that each electron exerts the same force on the coordinate, a coordinate which, in the absence of electrons, is acted upon by a harmonic oscillator potential. These conditions are what one would arrive at by making the very lowest order expansion of the energy, regarding lattice distortion as the small parameter. Then, as shown by Baraff *et al.* (1980), the energy to place the first electron on the defect is $\epsilon_L - E_{JT}$, where E_{JT} is the energy of lattice relaxation and ϵ_L is the energy which would have been needed if the lattice did not relax. The energy to put the second electron on is $\epsilon_L + U - 3E_{JT}$, where U is the added electrostatic repulsion. The significant physics here is that the amount of lattice displacement (as measured from the situation with no electrons in place) caused by adding two electrons is twice that of adding one. Therefore the energy lowering (also measured from the situation with no electrons in place) caused by two electrons is four times that of caused by one electron. The difference between energies to add the first and second electrons is then

$$\epsilon_2 - \epsilon_1 \equiv U_{\text{eff}} = U - 2E_{JT}.$$

It is this U_{eff} that is negative for the silicon vacancy.

The simple model here represents the total energy in all states of the system as being parabolas, much as virtually all configuration coordinate diagrams were constructed before 1985. Such simple diagrams can only be valid for relatively small displacements. When there is really large lattice distortion, the simple parabolas are of no use. One must instead use diagrams in which energies have been calculated in a serious manner. An early first effort in this direction was contributed by Baraff & Schluter (1985) who considered the total energy for various charge states of the gallium vacancy in GaAs. By motion of a single arsenic neighbour of the vacancy, this defect can be transformed into a nearest neighbour arsenic-vacancy–arsenic-antisite pair. The configuration coordinate on which the energy depended was taken to be the position of the displaced arsenic nearest neighbour atom. Since 1985, the technique for evaluating total energies has progressed to the extent that our current understanding of both DX and EL2 rests firmly on total energies and configuration coordinate diagrams as calculated by Chadi & Chang (1989) and by Dabrowski & Scheffler (1989).

(*e*) *Configuration dependent levels*

In so far as the energies depend on the configuration coordinates, one can also consider energy differences that depend on the configuration. Unlike the other levels, which were defined as the differences between minimum energies and were of use for describing statistical occupations, these levels are useful in reasoning about forces. This comes about as follows. The total energy depends on all the lattice coordinates, and so the force on the ith coordinate is the negative of the gradient of the energy with respect to that coordinate. Consider two forces on the same coordinate, R_i, one force arising in a system with j electrons and the other arising in a system containing $j-1$ electrons. That difference in force can be regarded as the force contributed by adding the jth electron to the system. It is given by the negative gradient of an energy difference. This makes it convenient to define this configuration dependent energy difference as a level:

$$\epsilon_j(Q) \equiv E_j(Q)-E_{j-1}(Q), \quad F_i = -\partial\epsilon_j(Q)/\partial R_i. \tag{9a, b}$$

Interestingly, this type of level is the simplest type to calculate. Apart from minor corrections, it is the eigenvalue of the Schrödinger equation that is solved in the Born–Oppenheimer calculation to get the wave functions and total energies at each configuration of the system. As an example of (9*b*), consider an effective mass level where a state is pulled down a small amount below the bottom of the conduction band by the coulomb attraction to an ionized donor. Such a level is insensitive to the exact location of lattice atoms in the neighbourhood of the donor. Its energy, on a configuration coordinate diagram, is independent of the coordinate. From (9*b*), that independence means that the electron in the state does not exert appreciable force on any of the lattice atoms. But this was already expected: the effective mass state is so spread out in space that its density is low and fairly uniform, so it cannot exert a force. Conversely, the energy levels of a deep level defect are found to be profoundly sensitive to the location of the atoms nearby, as has been shown in all defect calculations.

4. Examples of changing views on defects

Two examples of how progress in evaluating total energies accurately have been important in our understanding of large-lattice relaxation defects have already been mentioned: the DX centre and EL2. An account of how the theoretical calculations for these two defects has changed our fundamental ideas about defects is already in print (Baraff 1990). As this paper must be short there is only space to mention the existence of another article (Baraff 1992). This touches on another aspect of the effect (or lack of effect) of calculations. It cites a case in which a calculation of EL2 predating that of Dabrowski & Scheffler (1989) was ignored at the time it was presented because there was no confidence in the calculational method, even though, in retrospect, it turned out to be correct in concept and result. The EL2 story is still very much open as regards the microscopic identity of the defect. A summary of the present state of understanding appears in Baraff (1992).

New methods for evaluating the forces have emerged recently, and new methods for handling bigger and bigger systems have appeared. These have made it possible to determine the atomic arrangements of larger and larger numbers of atoms around the defects. The resulting accuracy has made it possible to learn more about defect properties in solids. These new methods are described in this volume.

References

Baraff, G. A. 1990 The importance of interaction between theory and experiment in understanding and identifying technologically important point defects. In *Defect control in semiconductors* (ed. K. Sumino) pp. 31–40. Amsterdam: North-Holland.

Baraff, G. A. 1992 Update: the mid-gap donor level EL2 in GaAs. In *Deep centers in semiconductors* (ed. S. T. Pantelides). New York: Gordon and Breach. (Revised version of Pantelides (1986).)

Baraff, G. A., Kane, E. O. & Schluter, M. 1980 Theory of the silicon vacancy: an Anderson negative-U system. *Phys. Rev.* B **22**, 5662–5686.

Baraff, G. A. & Schluter, M. 1979 New self-consistent approach to the electronic structure of localized defects in solids. *Phys. Rev.* B **19**, 4965–4979.

Baraff, G. A. & Schluter, M. 1985 Bistability and metastability of the gallium vacancy in GaAs: the actuator of EL2? *Phys. Rev. Lett.* **21**, 2340–2343.

Bernholc, J., Lipari, N. O. & Pantelides, S. T. 1980 Scattering-theoretic method for defects in semiconductors II: Self consistent formulation and application to the vacancy in silicon. *Phys. Rev.* B **21**, 3543–3562.

Car, P. & Parrinello, M. 1985 Unified approach for molecular dynamics and density functional theory. *Phys. Rev. Lett.* **55**, 2471–2474.

Chadi, D. J. & Chang, K. J. 1989 Energetics of DX formation in GaAs and $Al_x\,Ga_{1-x}As$ alloys. *Phys. Rev.* B **39**, 10063–10074.

Dabrowski, J. & Scheffler, M. 1989 Isolated arsenic–antisite defect in GaAs and the properties of EL2. *Phys. Rev.* B **40**, 10391–10401.

Kelly, P. J. & Car, R. 1992 Green's matrix calculation of total energies of defects in silicon. *Phys. Rev.* B **45**, 6543–6563.

Koster, G. F. & Slater, J. C. 1954 Simplified impurity calculation, *Phys. Rev.* **96**, 1208–1223.

Martin, G. & Makram-Ebeid, S. 1986 The mid-gap donor level EL2 in GaAs. In *Deep centers in semiconductors* (ed. S. T. Pantelides) p. 399. New York: Gordon and Breach.

Mooney, P. M. 1991 Donor related levels in GaAs and $Al_x\,Ga_{1-x}As$. *Semiconductor Sci. Technol.* **6** (10), B1–B8.

Newton, J. L., Chatterjee, A. P., Harris, R. D. & Watkins, G. D. 1983 Negative-U properties of the lattice vacancy in silicon. *Physica* B **116**, 219–223.

Pantelides, S. T. (ed.) 1986 *Deep centers in semiconductors.* New York: Gordon and Breach.

Stitch, I., Car, R., Parrinello, M. & Baroni, S. 1989 Conjugate gradient minimization of the energy functional: a new method for electronic structure calculations. *Phys. Rev.* B **39**, 4997–5004.

Watkins, G. D., Troxell, J. R. & Chatterjee, A. P. 1979 Vacancies and interstitials in silicon. In *Defects and radiation effects in semiconductors* (ed. J. H. Albany), pp. 16–30, Conf. Ser. N. 46. Bristol and London: Institute of Physics.

Ab initio Hartree–Fock treatment of ionic and semi-ionic compounds: state of the art

By R. Dovesi[1], C. Roetti[1], C. Freyria-Fava[1], E. Aprà[1], V. R. Saunders[2] and N. M. Harrison[2]

[1] *Department of Inorganic, Physical and Materials Chemistry, University of Torino, Via P. Giuria 5, I-10125 Torino, Italy*

[2] *SERC Daresbury Laboratory, Daresbury, Warrington WA4 4AD, U.K.*

The periodic *ab initio* Hartree–Fock approach is applied to the Li, Na, K, Be, Mg, Ca and Mn oxides, and to Al_2O_3 (corundum) and SiO_2 (α-quartz). A local basis set ('atomic orbitals') is used. The equilibrium geometry, the formation energy and the bulk modulus are calculated, with reasonable agreement with experiment. The influence of the environment on the oxygen ions is discussed through the Mulliken population and band structure data.

1. Introduction

The static and dynamic properties of ionic and semi-ionic crystals have usually been calculated by semi-classical methods based on Born-type semi-empirical formulae for the interatomic potential energy. An overview of the theory and of the different fields of application can be found in Catlow & Mackrodt (1982). The simplicity of the algorithm allows the investigation of complicated structures, including bulk defects and surface phenomena. On the other hand the nature of the method requires a careful analysis of the results when applied in conditions different from those in which the parameters have been determined. Obviously, no information concerning the wavefunction can be obtained from semi-classical techniques.

Recent advances in the speed and accuracy of the *ab initio* ('first principle') methods now permit the calculation of many ground state properties of simple ionic compounds. The *ab initio* methods can be considered at present as complementary tools to the semi-classical techniques, to which they can provide input data for the parameterization as an alternative to the experiment.

In this paper we present results obtained with our periodic *ab initio* Hartree–Fock CO (crystalline orbitals), LCAO (linear combination of atomic orbitals) program CRYSTAL (Pisani *et al.* 1988; Dovesi *et al.* 1992). In the past five years CRYSTAL has been applied to many ionic systems as well as to semiconductors (Causà *et al.* 1991), and molecular crystals (Dovesi *et al.* 1990) by different authors and at different levels of sophistication (basis set; numerical tolerances adopted in the calculation; use of all-electron or pseudopotential approaches). On the whole the agreement with experiment, when available, is satisfactory; on the other hand, to assess the quality of the results produced by a given method, it is important to refer to a relatively wide class of systems (and properties) investigated with basis sets of equivalent quality and at a comparable level of numerical accuracy. In this paper we perform such an analysis on a set of oxides: fully ionic (Li_2O, Na_2O, K_2O; BeO, MgO, CaO) as well as semi-ionic (corundum and α-quartz) systems will be considered; preliminary results

Phil. Trans. R. Soc. Lond. A (1992) **341**, 203–210
Printed in Great Britain
© 1992 The Royal Society

Table 1. *Basis sets used in the present study*

(The 'atomic orbitals' (AOS) are contractions of n gaussian type functions (GTFS); a GTF is the product of a gaussian times a polynomial in x, y, z. The AOS sharing the exponents of the GTFS are grouped in shells (s, sp and d shells). In the table the first contraction refers to an s shell, the others to sp shells. So, for example, 6-1 in the lithium case means that the basis consists of 5 AOS resulting from one s shell (contraction of six GTFS) and one (single gaussian) sp shell. The star means that d orbitals are included; in Mn two sets of d functions are used (4-1 contraction); in Ca and K contraction of three GTFS is used, while for Si, Al and O a single gaussian d shell is adopted. In the MnO case, small core (SC) pseudopotentials (PS) are used for Mn (Hay & Wadt 1985) and O (Durand & Barthelat 1975); in this case the first shell is sp. The exponents of the most diffuse gaussians are reported. In all cases the basis set has been optimized in the bulk.)

	cation				anion			
system		α^{sp}_{n-1}	α^{sp}_{n}	α^{d}		α^{sp}_{n-1}	α^{sp}_{n}	α^{d}
Li_2O	6-1	—	0.53	—	8-4-1-1	0.45	0.15	—
Na_2O	8-5-1-1	0.55	0.27	—	8-4-1-1	0.46	0.14	—
K_2O	8-6-5-1-1*	0.39	0.22	0.39	8-4-1-1	0.47	0.13	—
BeO	5-1-1	2.34	0.66	—	8-4-1-1	0.54	0.23	—
MgO	8-5-1-1	0.68	0.28	—	8-4-1-1	0.50	0.19	—
CaO	8-6-5-1-1*	0.45	0.25	0.38	8-4-1-1	0.50	0.17	—
Al_2O_3 (corundum)	6-2-1*	—	0.16	0.43	6-2-1*	—	0.35	0.8
SiO_2 (α-quartz)	6-2-1*	—	0.13	0.50	6-2-1*	—	0.36	0.8
MnO	PS–SC 3-1-1-1*	0.50	0.20	0.22	PS 4-1	—	0.22	—

for MnO will also be presented. Previous HF results on the above systems can be found in Dovesi *et al.* (1991), Lichanot *et al.* (1992), Causà *et al.* (1986), Silvi *et al.* (1992), Salasco *et al.* (1991) and Dovesi *et al.* (1987); references for the experimental data quoted in the following tables can be found in those papers, if not otherwise stated. The formation energy, equilibrium geometry, bulk modulus and elastic constants have been investigated; Mulliken population and band structure data will be used for characterizing the nature of the chemical bonds.

All but the MnO data can be reproduced with the version of the code distributed by the authors (Dovesi *et al.* 1992). It must be stressed that the HF–CO–LCAO method, as implemented in CRYSTAL, can be applied both to surface problems and to the study of neutral local defects with a supercell approach, with an accuracy similar to that documented below for the bulk properties. For reasons of space we will not report data referring to those applications.

2. The method and the basis set

We refer to previous papers for a discussion of the HF–CO–LCAO method. The scheme is affected by three different sources of error.

1. Numerical errors associated with the approximate treatment of the infinite Coulomb and exchange series and the reciprocal cell integration; numerical errors in the fitting procedure for obtaining the first and second derivatives of the energy. It can be shown that for the present compounds such errors are quite small (less than 2% for the elastic constants, which are second derivatives of the energy; negligible for the other properties).

Table 2. *Calculated and experimental equilibrium geometry and bulk modulus*

(a and c are the lattice parameters of unit cell and are given in ångstroms; x, y, z are fractionary coordinates of the indicated atom; the bulk modulus is in GPa; the underlined data have been extrapolated to the static limit, the others refer to room temperature. For the experimental sources, see text.)

		geometry parameters			bulk modulus	
system	space group	param.	calc.	exp.	calc.	exp.
Li_2O	Fm3m	a	4.57	4.57	93	89
Na_2O	Fm3m	a	5.48	5.55	58	—
K_2O	Fm3m	a	6.47	6.44	35	—
BeO	$P6_3mc$	a	2.68	2.68	253	214, 220
		c	4.34	4.35	—	224, 245
		x_O	0.380	0.377	—	—
MgO	Fm3m	a	4.20	4.19	186	167
CaO	Fm3m	a	4.87	4.79	128	120
Al_2O_3	$R\bar{3}c$	a	4.74	4.76	—	—
(corundum)		c	13.03	12.99	—	—
		z_{Al}	0.354	0.352	—	—
		z_O	0.304	0.306	—	—
		V	253.29	254.98	—	—
SiO_2	$P3_221$	a	4.93	4.92	—	34–37
(α-quartz)		c	5.42	5.40	—	—
		z_{Si}	0.470	0.470	—	—
		x_O	0.412	0.414	—	—
		y_O	0.269	0.267	—	—
		z_O	0.117	0.119	—	—
		V	114.13	113.12	—	—
MnO	Fm3m	a	4.53	4.44	152	147

2. Basis set limitations. The adopted basis set, as summarized in table 1, is probably close to a triple zeta basis set in a molecular context; as will be shown with reference to the properties of MgO (table 4), it can be assumed that the error associated with the use of the present basis sets is quite small for many of the properties investigated here.

3. The correlation error. It is well known that the Hartree–Fock method disregards the instantaneous interelectronic interactions. The largest part of the difference between the calculated and experimental data to be discussed below is to be attributed to such an approximation. Methods exist for overcoming the limitations of the HF approach, which are common practice in molecular quantum chemistry (Wilson 1984). No standard implementation of these techniques for periodic systems exists at present. A simple correction of the HF total energy is possible, however, through correlation only density-functional schemes (Colle & Salvetti 1975; Perdew 1986, 1991), that have been shown to perform quite well for semiconductors (Causà *et al.* 1991); such a correction has been applied in the present paper in the evaluation of the HF formation energy.

Table 3. *Experimental and calculated binding energies* (hartree)

(The correlation corrections are evaluated according to the Colle & Salvetti (1975 (CS)) and Perdew's (1986 (P86) and 1991 (P91)) functionals. The numbers in parenthesis are the percentage errors with respect to the experiment.)

system	HF	HF+CS	HF+P86	HF+P91	exp.
Li_2O	0.300	0.403	0.410	0.399	0.439
	(−32)	(−8)	(−7)	(−9)	
Na_2O	0.189	0.277	0.286	0.276	0.331
	(−44)	(−16)	(−14)	(−17)	
K_2O	0.138	0.224	0.249	0.234	0.293
	(−54)	(−24)	(−15)	(−20)	
BeO	0.339	0.416	0.432	0.426	0.446
	(−24)	(−7)	(−3)	(−4)	
MgO	0.269	0.342	0.361	0.356	0.377
	(−29)	(−9)	(−4)	(−6)	
CaO	0.279	0.357	0.376	0.371	0.399
	(−30)	(−10)	(−6)	(−7)	
Al_2O_3	0.894	1.124	1.174	1.159	1.161
	(−23)	(−3)	(+1)	(0)	
SiO_2	0.512	0.649	0.675	0.667	0.704
	(−27)	(−8)	(−4)	(−5)	
MnO	0.232	0.318	0.344	0.338	0.343
	(−32)	(−7)	(0)	(−1)	

3. Results and discussion

In table 2 the geometry optimized with the basis sets of table 1 are reported and compared with the experimental data; where possible, the latter have been corrected for the zero point and temperature effects (static limit). It turns out that the error is negligible for Li_2O, BeO, MgO; below 1% for SiO_2, Al_2O_3 and Na_2O (although we note that we were not able to find thermal expansion data for the latter system and for K_2O, for which the temperature effects are expected to be quite large; using the thermal expansion data of NaF and KF, an athermal limit of 5.49 Å† and 6.38 Å is obtained for the two systems); between 1 and 2% for CaO and K_2O; in the case of MnO the error is about 2%. The error increases along the series Li, Na, K and Be, Mg, Ca, because of the increasing relative importance of dispersion and intraionic correlation effects with respect to the Coulomb and short-range repulsion terms. In extreme situations, where large cations are involved, as for example with rubidium or silver halides, the error can be as large as 10% (Aprà *et al.* 1991). The last two columns of table 2 report the calculated and experimental bulk moduli. The HF results are always larger by about 10% than the experimental data, in line with molecular experience (Hehre *et al.* 1986).

In table 3 the binding energy (BE) evaluated with respect to the isolated atoms is reported. The HF BE is 20 to 50% smaller than the experimental one, due to the large correlation contribution to the bond formation. The third, fourth and fifth columns show the binding energy obtained after adding *a posteriori* correlation contributions to the HF energy, by integrating the HF charge density with the formulae proposed by Colle & Salvetti (1975) and Perdew (1986, 1991). The three functionals appear to perform roughly equivalently; the HF error is always reduced; the final BE remain,

† $1\ \text{Å} = 10^{-10}\ \text{m} = 10^{-1}\ \text{nm}$.

Table 4. *Basis set effects on bulk properties of* MgO *in the rocksalt structure*

(Basis set (*c*) is described in table 1. (*d*) and (*e*) are obtained by adding a single gaussian d-shell to oxygen and magnesium respectively ($\alpha = 0.65$ bohr^{-2} in both cases). (*b*) is obtained from (*c*) by contracting one more gaussian into the valence inner shell of magnesium and reoptimizing the outer gaussian shell ($\alpha = 0.40$ bohr^{-2}); (*a*) is obtained from (*b*) performing a similar contraction on oxygen ($\alpha = 0.21$ bohr^{-2}). t_1 and t_2 are the CPU times (in CRAY-YMP seconds) required by the integral and SCF part for the calculation of a single energy point. E_{tot} and BE (in hartrees) are the total and binding energies; a_0 (in ångströms) is the lattice parameter; B, C_{11}, C_{12} and C_{44} (in GPa) are the bulk modulus and the elastic constants. The number of atomic orbitals per unit cell is 18, 22, 26, 31 and 36 from (*a*) to (*e*). The experimental data at the athermal limit are 314, 94, 160 GPa for C_{11}, C_{12} and C_{44} respectively (Sumino *et al.* 1983).)

	basis set										
	cation	anion	t_1	t_2	E_{tot}	BE	a_0	B	C_{11}	C_{12}	C_{44}
(*a*)	8-61	8-51	32	2	−274.6724	0.267	4.193	195	384	100	199
(*b*)	8-61	8-411	39	4	−274.6741	0.268	4.201	186	357	100	190
(*c*)	8-511	8-411	67	5	−274.6748	0.269	4.205	181	340	101	185
(*d*)	8-511	8-411*	98	7	−274.6752	0.270	4.205	182	333	105	184
(*e*)	8-511*	8-411*	144	9	−274.6799	0.274	4.195	183	326	111	183

however, in all cases smaller than the experimental result (Lide 1991) by something between 0 and 20%. A comparison with the results of a similar study on semiconductors (seventeen III–V systems (Causà *et al.* 1991)) seems to indicate that for ionic compounds the *a posteriori* DF correction is less effective than for covalent systems: in the former case the error (after the P86 correction) was nearly independent of atomic number, and the mean error was 2.5%, whereas in the present case it is greater than 4%.

In table 4 the effect of the basis set on the static properties of MgO is documented. Basis set (*a*) is relatively poor, although more flexible than a minimal basis set. At the other extreme, basis set (*e*) is quite complete, containing three valence sp shells on oxygen and two valence sp shells on magnesium; a set of polarization functions is included on both atoms. If we take as a reference basis set (*a*), the table shows that it is important to provide oxygen additional variational freedom for allocating the two extra electrons coming from magnesium; from (*b*) on, however, the properties remain essentially constant as the basis set is enlarged. The exceptions are the C_{11} and C_{12} elastic constants; the important role of d functions, documented in the table, is related to the polarization of both the cation and of the anion, allowed by the reduced symmetry of the atomic site in the deformed unit cell. We note that the large Cauchy violation exhibited by MgO is well reproduced by the HF theory.

Figure 1 offers another example of application of the HF–LCAO technique; CaO undergoes a B1 ('NaCl structure') to B2 ('CsCl structure') phase transition at about 60–70 GPa (Mammone *et al.* 1981). Theoretically, such a transition has been studied, among others, by Mehl & Cohen (1988) with an LAPW technique (transition pressure 54 GPa); previous calculations with the CRYSTAL program performed by using pseudopotentials (D'Arco *et al.* 1992; 68 GPa) or smaller basis sets (D. M. Sherman, personal communication; 74 GPa) produced a transition pressure quite close to the present one (70 GPa) and in reasonable agreement with experiment.

In table 5 the Mulliken population data for the systems investigated are reported. The oxygen net charge is very close to $-2|e|$ for the Li, Na, K and Mg compounds; around $-1.8|e|$ for CaO and BeO and $-1.5|e|$ for MnO. It is $-1.2|e|$ and $-1.0|e|$ in

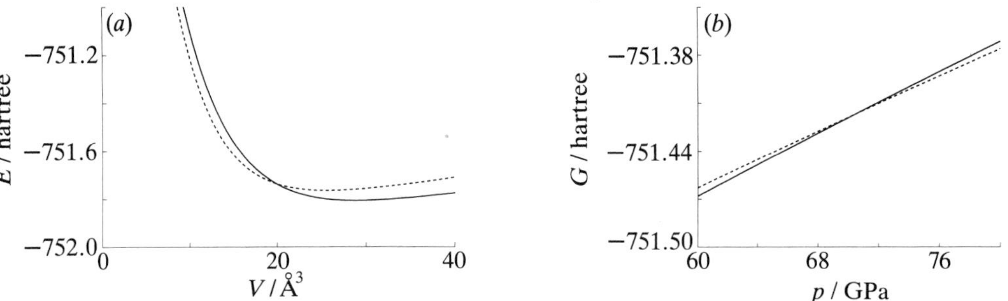

Figure 1. (*a*) Total energy of CaO in the B1 and B2 phases as a function of the unit cell volume. (*b*) Calculated free energy against pressure curves for CaO in the B1 and B2 phases. ———, B1; - - - - -, B2.

Table 5. *Mulliken population analysis data*

(q(O) is the oxygen net charge; B(i–j) is the bond population; $r_O(r_x)$ is the square root of the anion (cation) spherical second-order moment; the electronic charge density is partitioned according to a Mulliken analysis. N is the coordination number of the oxygen atom; R(X–O) and R(O–O) are the shortest distances between the indicated atoms. Distances in bohrs.)

system	q(O)	q(X)	B(X–O)	B(O–O)	r_O	r_X	N	R(X–O)	R(O–O)
Li_2O	−1.937	0.969	−0.007	−0.020	4.42	0.98	8	3.74	6.11
Na_2O	−2.067	1.034	−0.028	−0.003	4.83	2.45	8	4.54	7.42
K_2O	−2.014	1.007	−0.050	−0.001	4.79	4.33	8	5.27	8.60
BeO	−1.746	1.746	0.043	−0.017	4.07	1.07	4	3.09	5.04
MgO B1	−1.947	1.947	−0.004	−0.025	4.32	2.24	6	3.97	5.61
MgO B2	−1.824	1.824	0.012	−0.096	4.22	2.42	8	4.34	5.01
CaO B1	−1.833	1.833	−0.018	−0.001	4.40	4.00	6	4.60	6.50
CaO B2	−1.877	1.877	−0.020	−0.051	4.33	3.95	8	4.83	5.57
Al_2O_3	−1.220	1.830	0.148	−0.020	3.84	3.18	3	3.50	4.77
			0.107				3	3.72	
SiO_2	−1.018	2.036	0.290	−0.010	3.73	3.76	4	3.04	4.94
MnO	−1.465	1.465	0.034	−0.008	4.16	4.45	6	4.31	6.10

corundum and α-quartz respectively. In the BeO case the 0.2 electron difference with respect to fully ionic bonding is due to the important participation of the p orbitals of the cation; a similar role is played by the d orbitals of Ca in CaO (0.18 electrons). The B(X–O) column provides information for the characterization of the bonds; negative values indicate that the interaction between the electrons of the cation and the anion is repulsive (short-range repulsive or 'exclusion' forces), as expected for a fully ionic compound; it is to be noted that also from this point of view BeO is 'anomalous' with respect to the other simple oxides (a small positive bond population). The large positive values for corundum and α-quartz are a measure of the covalent character of the X–O bond. According to the B(X–O) values α-quartz is much more covalent than corundum, as expected. Further information can be obtained by B(O–O), which is a measure of the repulsion between the oxygens; in the alkali series, for example, it is much larger in Li_2O than in Na_2O and K_2O, indicating that the equilibrium configuration in the former case is mainly determined by the oxygen–oxygen repulsion; a similar trend is shown by the Be, Mg and Ca series. The comparison of the B2 data for MgO (referring to the equilibrium volume of the B1 phase) and CaO (referring to the equilibrium volume of the B2 phase) shows that the energetics of this

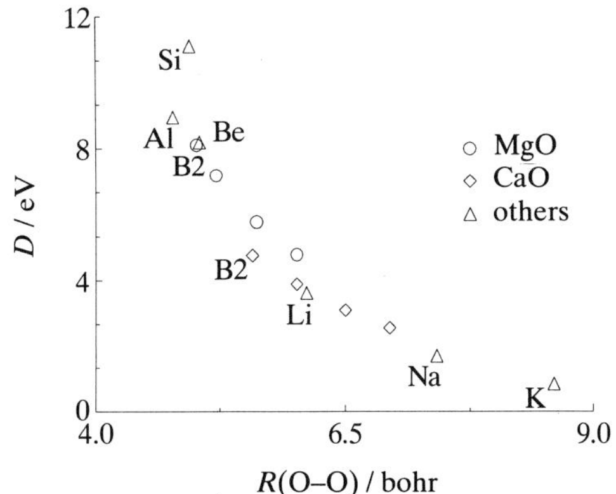

Figure 2. Correlation between the width of the p band of oxygens (eV) and the distance R_{O-O}(Å) between the nearest neighbours oxygen atoms. The three values for MgO refer to $a_0 = 3.90$, 4.20 (equilibrium) and 4.50 Å; for CaO to 4.20, 4.87 (equilibrium) and 5.20 Å. Two points (one for MgO and one for CaO) referring to the B2 (CsCl) structure are also reported. For the other oxides only the cation is indicated.

more closely packed structure are essentially determined by the O–O repulsion. The r_O and r_X columns compare the calculated radii of the ions (square root of the spherical second-order moment evaluated after partitioning the electronic charge of the unit cell according to Mulliken). It turns out that the size of the oxygen atom is quite different for ionic and covalent compounds; data not reported in the table show that r_O exhibits considerable change when the system is compressed, supporting the need of 'breathing' terms in semi-classical models of the oxides (Schroeder 1966).

The density of states of the ionic compounds is very simple: the bands are flat, separated by large gaps; each band can be characterized in terms of a particular atomic orbital of the atoms of the unit cell; in the case of the oxides, the highest valence bands are always oxygen (non-bonding) p bands; the participation of the cation to those bands is negligible; the 'dispersion' of the band (and thus the bandwidth) is a measure of the interaction between the oxygens. Figure 2 correlates the width of the oxygen p bands with R(O–O), the distance between first neighbours oxygens (the coordination of the oxygen atom is given in table 5). The figure shows that a similar correlation exists for the compounds investigated with the exception only of SiO_2, which appears to belong to a different 'family' due to its strong covalent character.

4. Conclusions

The Hartree–Fock LCAO technique, as implemented in CRYSTAL, has been shown to provide reasonable results and useful interpretation schemes in the study of ionic and semi-ionic compounds. Given its availability from the authors and the relatively low cost of the calculations, CRYSTAL can be considered a standard tool in the study of the electronic structure of simple ionics, and a useful complement to semi-classical methods.

References

Aprà, E., Stefanovich, E., Dovesi, R. & Roetti, C. 1991 An *ab initio* Hartree–Fock study of silver chloride. *Chem. Phys. Lett.* **186**, 329–335.

Catlow, C. R. A. & Mackrodt, W. C. 1982 Computer simulation of solids. *Lecture notes in physics*, vol. 166. Berlin: Springer Verlag.

Causà, M., Dovesi, R., Pisani, C. & Roetti, C. 1986 Electronic structure and stability of different crystal phases of magnesium oxide. *Phys. Rev.* B **33**, 1308–1316.

Causà, M., Dovesi, R. & Roetti, C. 1991 Pseudopotential Hartree–Fock study of seventeen III–V and IV–IV semiconductors. *Phys. Rev.* B **43**, 11937–11943.

Colle, R. & Salvetti, O. 1975 Approximate calculation of the correlation energy for the closed shells. *Theor. Chim. Acta* **37**, 329–334.

D'Arco, Ph., Jolly, L. H. & Silvi, B. 1991 Periodic Hartree–Fock study of topochemical reactions: phase transition in CaO. In *Physics of the earth and planetary interiors*, vol. 71.

Dovesi, R., Pisani, C., Roetti, C. & Silvi, B. 1987 The electronic structure of α-quartz: a periodic Hartree–Fock calculation. *J. chem. Phys.* **86**, 6967–6971.

Dovesi, R., Causà, M., Orlando, R., Roetti, C. & Saunders, V. R. 1990 *Ab initio* approach to molecular crystals: a periodic Hartree–Fock study of crystalline urea. *J. chem. Phys.* **92**, 7402–7411.

Dovesi, R., Roetti, C., Freyria-Fava, C., Prencipe, M. & Saunders, V. R. 1991 On the elastic properties of lithium, sodium and potassium oxide. An *ab initio* study. *Chem. Phys.* **156**, 11–19.

Dovesi, R., Saunders, V. R. & Roetti, C. 1992 *CRYSTAL92. User documentation.* University of Torino and SERC Daresbury Laboratory.

Durand, P. & Barthelat, J. C. 1974 New atomic pseudopotentials for electronic structure calculations of molecules and solids. *Chem. Phys. Lett.* **27**, 191–194.

Hay, P. J. & Wadt, W. R. 1985 *Ab initio* effective core potentials for molecular calculations. Potentials for the transition metal atoms Sc to Hg. *J. chem. Phys.* **82**, 270–310.

Hehre, W. J., Radom, L., Schleyer, P. V. R. & Pople, J. A. 1986 *Ab initio* molecular orbital theory. New York: Wiley Interscience.

Lichanot, A., Chaillet, M., Larrieu, C., Dovesi, R. & Pisani, C. 1992 *Ab initio* Hartree–Fock study of beryllium oxide: structure and electronic properties. *Chem. Phys.* (In the press.)

Lide, D. R. (ed.) 1991 *Handbook of chemistry and physics*, 72nd edn. CRC Press.

Mammone, J. F., Mao, H. K. & Bell, P. M. 1981 Equations of state of CaO under static pressure conditions. *Geophys. Res. Lett.* **8**, 140–142.

Mehl, M. J. & Cohen, R. E. 1988 Linearized augmented plane wave electronic structure calculations of MgO and CaO. *J. geophys. Res.* **93**, 8009–8022.

Perdew, J. P. 1986 Density-functional approximation for the correlation energy of the inhomogenous electron gas. *Phys. Rev.* B **33**, 8822–8824. (Erratum: *Phys. Rev.* B **34**, 7406.)

Perdew, J. P. 1991 In *Electronic structure of solids* (ed. P. Ziesche & H. Eschrig). Berlin: Akademie Verlag.

Pisani, C., Dovesi, R. & Roetti, C. 1988 Hartree–Fock *ab initio* treatment of crystalline systems. *Lecture notes in chemistry*, vol. 48. Heidelberg: Springer Verlag.

Salasco, L., Dovesi, R., Orlando, R., Causà, M. & Saunders, V. R. 1991 A periodic *ab initio* extended basis set study of α-Al_2O_3. *Molec. Phys.* **72**, 267–277.

Schroeder, U. 1966 A new model for lattice dynamics ('breathing shell model'). *Solid State Commun.* **4**, 347–349.

Silvi, B., D'Arco, Ph., Saunders, V. R. & Dovesi, R. 1991 Periodic Hartree–Fock study of minerals: tetracoordinated silica polymorphs. *Phys. Chem. Minerals* **17**, 674–680.

Sumino, Y., Anderson, L. O. & Susuki, I. 1983 Temperature coefficients of elastic constants of single crystal MgO between 80 and 1300 K. *Phys. Chem. Minerals* **9**, 38–47.

Wilson, S. 1984 *Electron correlation in molecules.* Oxford: Clarendon Press.

Role of parallel architectures in periodic boundary calculations

By M. C. Payne[1], L. J. Clarke[2] and I. Stich[1,3]

[1] *Cavendish Laboratory, Madingley Road, Cambridge CB3 0HE, U.K.*

[2] *Edinburgh Parallel Computer Centre, University of Edinburgh, Mayfield Road, Edinburgh EH9 3JZ, U.K.*

[3] *Institute of Inorganic Chemistry of the Slovak Academy of Science, Bratislava, Czechoslovakia*

Ab initio computations can be used to determine the values of a wide variety of physical properties of atoms, molecules and solids. The calculations are computationally demanding and can only be applied to small systems. One possible method for overcoming this limitation is to use parallel computers which, in principle, can provide unlimited computational power. In this paper the technical difficulties associated with parallelizing *ab initio* calculations are reviewed and a case study detailing the implementation of total energy pseudopotential codes on parallel machines is presented.

1. Introduction

There are a large number of *ab initio* or first principles methods available for calculating the physical properties of atoms, molecules and solids. These methods can be used to calculate the values of a wide range of physical properties and the results are consistently found to be in very good agreement with experimental values. The success of these methods has been so great that they are now generally regarded as predictive. However, all these methods are computationally demanding and even calculations for small systems, containing only a few atoms, can only be performed on powerful workstations or supercomputers. At the same time it is increasingly apparent that many complex processes in physics, chemistry, biology and materials science can only be solved with *ab initio* modelling of systems a hundred or thousand times larger than those now studied. There are two ways of increasing the size of system that is accessible to *ab initio* modelling. One is to improve the algorithms and hence increase the computational efficiency of the method. The other is to use a more powerful computer. It is now accepted that the only way to achieve significant increases in computational power in the immediate future is by harnessing together the power of many processors in a parallel computer. The most powerful parallel machines presently available achieve computational performances of the order of 10 GFlop but it is clear that within the next five years at least one teraflop machine will become available.

Although the number of processors in a parallel machine can, in principle, be increased without limit it is not clear what ultimately limits the technology that links the processors together into a useful machine and so the ultimate performance that can be achieved from such machines is not known. However, it is obvious that computational power of the order of ten thousand times greater than present

Phil. Trans. R. Soc. Lond. A (1992) **341**, 211–220
Printed in Great Britain
© 1992 The Royal Society

supercomputers and one hundred thousand times greater than present workstations will be available. These figures are so large that it appears that the only obstacle to *ab initio* modelling of complex processes is the implementation of existing techniques on parallel machines. Unfortunately this is not the case because of the severe scaling of the computational time with system size of most *ab initio* techniques. If the computational time scales as the fourth power of the system size then the 10000 fold increase in computational power yields only a ten fold increase in the size of the system that can be studied; still one or two orders of magnitude removed from the target figure for complex processes. This is not to say that the *ab initio* method cannot be applied to a large range of important scientific problems. However, it is clear that implementation of *ab initio* codes on parallel computers alone cannot be viewed as the solution to the limitations on system size. Another problem that is particularly acute in many *ab initio* methods is that it may not be possible to implement the current numerical algorithms on parallel machines. One example is the QR factorization method for matrix diagonalization (Wilkinson 1965). The computational time for the QR method scales as N^3, where N is the size of the matrix, but this algorithm cannot be implemented on a parallel computer. Other matrix diagonalization techniques such as the Jacobi method (Wilkinson 1965) can be implemented on parallel machines. However, the computational cost of these methods scales as N^4. Hence, in the case of matrix diagonalization the additional power of a parallel computer would be completely negated by the reduction in efficiency of the numerical algorithm. To achieve a real gain from the use of parallel computers it is crucial to develop numerical methods that have the lowest possible scaling of computational cost with system size. Only after this has been achieved will an increase in computational power yield a significant increase in the size of system accessible to the *ab initio* technique.

In the remainder of this paper we concentrate on just one *ab initio* modelling method, the total energy pseudopotential method. In §2 it will be shown how a number of algorithmic developments in recent years have drastically improved the computational efficiency of this method. The implementation of total energy pseudopotential calculations on parallel machines will be described in §3 and prospects for the future are briefly summarized in §4.

2. The total energy pseudopotential method

A number of reviews (Cohen 1984; Joannopoulos 1985; Pickett 1989) provide an overview of the total energy pseudopotential method. Technical details can be found in Ihm *et al.* (1979) and Denteneer & van Haeringen (1985). Recent algorithmic developments are described in detail in Payne *et al.* (1992). The elements of a total energy pseudopotential calculations are as follows: the ions are represented by pseudopotentials (Phillips 1958; Heine & Cohen 1970) so that only the wavefunctions of the valence electrons are included in the calculation; density functional theory (Hohenberg & Kohn 1964; Kohn & Sham 1965) is used to represent the effects of electron–electron interactions; calculations are performed on a periodically repeated unit cell (referred to as a supercell) so that the electron wavefunctions at each k-point in the Brillouin zone can be expanded in terms of a discrete plane wave basis set as follows

$$\psi_{n,\boldsymbol{k}} = \sum_{\boldsymbol{G}} c_{\boldsymbol{G}} \exp(\mathrm{i}[\boldsymbol{k}+\boldsymbol{G}]\cdot\boldsymbol{r}), \tag{1}$$

where $\boldsymbol{k}$ is the k-point in the Brillouin zone and the summation is over reciprocal lattice vectors. This basis set has to be truncated and the convenient choice is to include all basis states for which $|\boldsymbol{k}+\boldsymbol{G}|$ is less than a cut-off wavenumber, G_{cut}.

The electronic states are given by the self-consistent solutions of the Kohn–Sham equations (Kohn & Sham 1965)

$$(-(\hbar^2/2m)\nabla^2 + V_{\text{ion}}(\boldsymbol{r}) + V_{\text{H}}(\boldsymbol{r}) + V_{\text{XC}}(\boldsymbol{r}))\,\psi_{n,\boldsymbol{k}} = \epsilon_n\,\psi_{n,\boldsymbol{k}}, \tag{2}$$

where V_{ion} is the ionic potential, V_{H} is the Hartree potential, V_{XC} is the exchange-correlation potential and ϵ_n is the Kohn–Sham eigenvalue. Self-consistent solutions are required because the Hartree and exchange-correlation potentials depend on the electronic density, $n(\boldsymbol{r})$.

The earliest implementations of the total energy pseudopotential method used standard matrix diagonalization techniques to obtain the Kohn–Sham eigenstates at a computational cost proportional to N_{PW}^3, where N_{PW} is the number of plane wave basis states used to expand the electronic wavefunctions. This appears to be a reasonable scaling of the computational time with system size. However, the number of plane wave basis states required is at least one hundred times larger than the number of atoms in the unit cell. Restrictions on computer memory and computational time limited such calculations to the order of a few tens of atoms in the unit cell. Furthermore, it is harder to achieve self-consistency as the size of the system increases so that the true scaling of computational time with system size is more accurately represented by the fourth or higher power of the number of atoms in the unit cell. Since only the few occupied Kohn–Sham eigenstates are required to compute the total energy it is clear that a significant saving in computational time can be achieved by using iterative matrix diagonalization techniques and including only the occupied electronic states in the calculation. A number of such methods have been developed but the most significant breakthrough in total energy calculations was achieved by Car & Parrinello (1985). They developed the molecular dynamics method for performing total energy pseudopotential calculations which included several novel techniques for increasing the efficiency of the calculations. In addition to the use of iterative matrix diagonalization techniques the process of iterating to self-consistency was overlapped with the process of determining the Kohn–Sham eigenstates thus decreasing the additional computational cost of achieving self-consistency. The basic operation common to all iterative matrix diagonalization methods involves the multiplication of the trial wavefunction by the hamiltonian matrix. Car & Parrinello significantly increased the speed of this multiplication by dividing the operations between real and reciprocal space. The product of the Kohn–Sham hamiltonian H and the trial wavefunction is given by

$$H\psi_{n,\boldsymbol{k}} = -(\hbar^2/2m)\,|\boldsymbol{k}+\boldsymbol{G}|^2\psi_{n,\boldsymbol{k}}(\boldsymbol{G}) + (V_{\text{ion}}(\boldsymbol{r}) + V_{\text{H}}(\boldsymbol{r}) + V_{\text{XC}}(\boldsymbol{r}))\,\psi_{n,\boldsymbol{k}}(\boldsymbol{r}), \tag{3}$$

where the first term on the right-hand side is the product of the wavefunction and the kinetic energy operator, which is diagonal in reciprocal space, and the second term is the product of the wavefunction and the potential energy operator, which is diagonal in real space if the ions are represented by local pseudopotentials. It requires N_{PW}^2 operations to evaluate the product of a matrix and a vector but by rewriting this product as shown in (2) this cost is reduced to N_{PW} operations for the kinetic energy operator and $16N_{\text{PW}}$ operations for the potential energy operator. The factor of 16 arises because the Fourier transform mesh must be larger than the wavefunction array to avoid 'wrapround' error in the calculation. A further significant improvement is that the hamiltonian matrix can be stored in $17N_{PW}$ words of memory by dividing it between real and reciprocal space in contrast to the N_{PW}^2 words of memory required to store the complete matrix in a single space. However, to exploit this reduction in computational cost the wavefunction must be transformed from reciprocal space to real space and the product of the wavefunction and the potential energy operator must be transformed from real space to reciprocal space. These transformations can be performed in $16N_{PW}\ln(16N_{PW})$ operations using fast Fourier transform techniques.

A number of other algorithms have been developed for performing total energy pseudopotential calculations, most notably conjugate gradients techniques that directly minimise the Kohn–Sham energy functional (Gillan 1989; Teter *et al.* 1989). These methods converge the electronic configuration to its groundstate faster than the molecular dynamics method but exploit all the features of the original method. In all these methods there are two operations that dominate the computational cost: the Fourier transforms and the operation of orthogonalizing the electronic wavefunctions. If there are N_B occupied electronic bands, which is typically of the order of $0.01N_{PW}$, then the total cost of the Fourier transforms scales as $16N_B N_{PW}\ln(16N_{PW})$ and the cost of orthogonalizing the wavefunctions scales as $N_B^2 N_{PW}$. Therefore the cost of the Fourier transforms dominates for small systems and the cost of orthogonalizing the electronic wavefunctions dominates for large systems yielding a computational time that scales as the cube of the size of the system for large systems. The cost of implementing non-local pseudopotentials used to dominate the computational time for large systems, typically the cost was a factor of 10 greater than the cost of that orthogonalizing the electronic wavefunctions. Recently a method has been developed for implementing non-local pseudopotentials in which the computational time scales as $N_B N_{PW}$ (King-Smith *et al.* 1991) so this operation no longer dominates the computational cost for large systems. Using any of the recently developed methods total energy pseudopotential calculations can now be routinely performed for systems containing up to 100 atoms in the unit cell on workstations and conventional supercomputers.

It has only been possible to provide the briefest overview of recent algorithmic developments in the total energy pseudopotential technique. The implementation of these methods on parallel computers is described in the following section. It is remarkable that the original matrix diagonalization method used for total energy pseudopotential calculations could not be implemented efficiently on a parallel computer but that recently developed algorithms are not only much more efficient but are very well suited to implementation on parallel machines.

3. Implementation of total energy pseudopotential codes on parallel computers

The central quantities in a total energy pseudopotential calculation are the electronic wavefunctions which can be represented by a complex array of the form

$$\Psi(N_{PW}, N_B, N_K), \tag{4}$$

where N_K is the number of k-points used for Brillouin zone sampling and N_{PW} and N_B have been defined above.

The structure of the wavefunction array suggests several methods for dividing the data across a parallel machine, for instance by k-point or by band or by plane waves.

To determine which of these strategies is possible for large systems we must consider the variation of N_{PW}, N_B and N_K with the number of atoms in the unit cell, N_A. These are:

$$\left.\begin{aligned} &N_{PW} \approx (100\text{–}1000)N_A, \quad N_B \approx N_A, \\ &N_K \approx \begin{cases} 100/N_A, & N_A \leqslant 100, \\ 1, & N_A \geqslant 100. \end{cases} \end{aligned}\right\} \tag{5}$$

It is obvious from these scalings that parallelization by k-point is not possible for large systems. However, it does appear possible to parallelize by band. If this strategy is adopted it is necessary to store the hamiltonian 'matrix' and a number of other arrays of similar size on each compute node. To store these arrays a MWord of memory per node would be needed for systems containing of the order of 100 atoms in the unit cell and calculations for larger and larger systems would require more and more memory per compute node. This is uneconomic and hence parallelization by band cannot be adopted. There is no alternative but to distribute the plane wave basis states of each wavefunction over the machine. To determine the most efficient choice for the distribution it is necessary to consider the operations that are performed on the wavefunctions. As described in the previous section, the efficiency of modern algorithms relies on the division of operations between real and reciprocal space and the use of fast Fourier transform (FFT) techniques to transform between the two spaces. Unfortunately, the FFT is a highly non-local operation and hence it places extreme demands on the communications system of the parallel computer. Apart from the FFT all the operations required to perform a total energy calculation are local (or localized in the case of the most efficient implementation of non-local pseudopotentials) in either real or reciprocal space. The actual distribution of the data is irrelevant provided that the distribution of the electronic wavefunctions and the hamiltonian in each space are identical. In this case, local operations require no communication between nodes. Therefore, the distribution of the wavefunctions is determined purely by the requirements of the FFT. Since the properties of each individual machine will determine the most efficient distribution of the wavefunctions we can only proceed by considering specific examples. We consider only machines that are now available on which total energy pseudopotential codes have been successfully implemented.

The first example considered is the Connection Machine (Brommer *et al.* 1992). The Connection Machine is an example of a massively parallel computer that contains a very large number of compute nodes of modest performance. Although the performance of a single processor may be only a fraction of 1 MFlop the combination of tens of thousands of processors yields a theoretical performance in excess of 10 GFlop. The Connection Machine is a particularly interesting example as it was designed to perform FFTs particularly efficiently and thus the connectivity of the communications system is well suited to this use. A library of microcoded FFT subroutines exists for performing FFTs and the distribution of the wavefunctions is dictated by these routines. Implementation of total energy pseudopotential codes on the Connection Machine can be achieved by rewriting code in FORTRAN 90 using vector oriented Fortran 90 statements to ensure that relevant operations are performed in parallel across all the nodes of the machine and calling the FFT library where appropriate.

The other class of machine that will be considered is typified by the machines manufactured by Intel and Meiko, which consist of a relatively modest number of

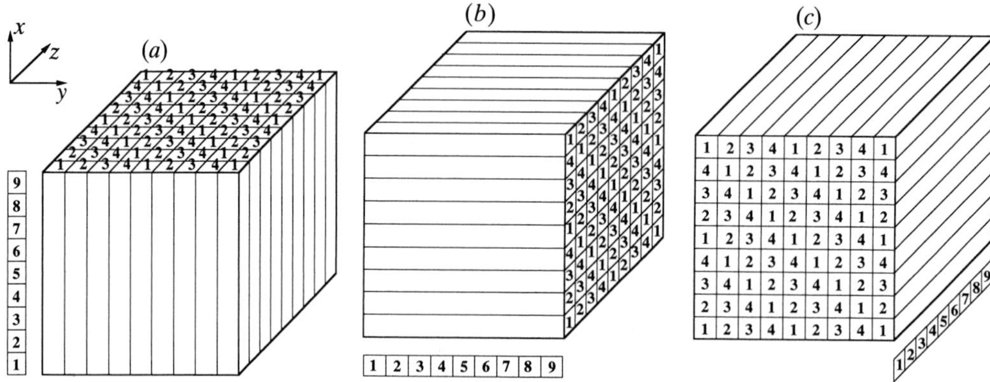

Figure 1. Illustration of the distribution of data over processors for a sequence of three one-dimensional Fourier transforms in (*a*) x, (*b*) y and (*c*) z directions required to perform a three-dimensional Fourier transform. The figure shows a $9 \times 9 \times 9$ Fourier transform performed on a machine with four compute nodes. The numbers in boxes (1–9) label the points of the Fourier transform and the node indices (1–4) show which processor performs the one-dimensional transform.

powerful compute nodes. In both cases the compute nodes are Intel i860 processors capable of a performance in excess of 40 MFlop. There are no fully distributed three-dimensional FFT routines available on these machines and the majority of the effort involved in implementing total energy codes on these machines involves writing programs to perform the FFT (Clarke *et al.* 1992). However, this does provide the freedom to choose a distribution of data that minimizes the communications requirement of the FFT. All multi-dimensional FFTs can be performed as a sequence of one-dimensional transforms. Each one-dimensional transform requires a large exchange of data but this can be avoided by ensuring that the data for each transform lies on the same compute node. Therefore, a distribution of data by columns along the direction of the transform ensures that each one-dimensional transform can be performed without inter-processor communication. However, a global exchange of data is required between each set of one-dimensional transforms to redistribute the data by columns along the next transform direction. The distribution of data for a transformation from a reciprocal space to real space is illustrated schematically in figure 1. The data is initially distributed over the processors by columns along the x direction, as shown in figure 1*a*. The first set of one-dimensional Fourier transforms are performed along the x direction. This is followed by the first global exchange of data between processors so that the data is now arranged by columns along the y direction, as illustrated in figure 1*b*. The one-dimensional transforms along the y direction are then carried out. A second global exchange of data is then performed so that the data is finally arranged by columns along the z direction, as illustrated in figure 1*c*. The final set of one-dimensional transforms along the z direction are then performed. At the end of this sequence of steps the transformation from reciprocal space to real space is complete.

If the communications performance of the machine is relatively poor and the number of processors is not greater than the size of the largest one-dimensional Fourier transform size a significant increase in efficiency can be achieved by combining the final two steps of the above sequence by distributing the data by yz planes, as illustrated in figure 2. The first step of the three-dimensional Fourier

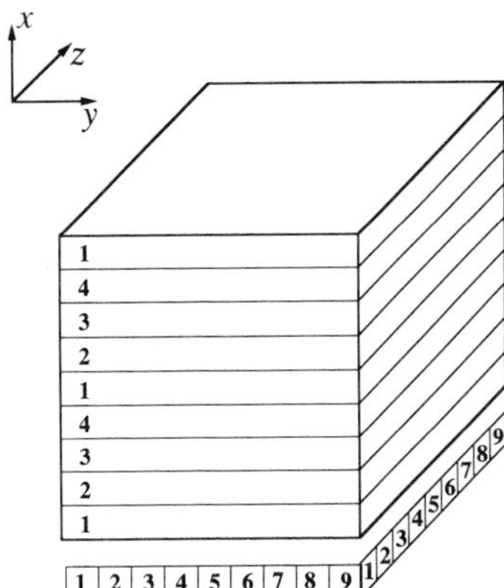

Figure 2. Distribution of data over nodes to replace the transforms shown in figure 1*b*, *c*. The data is distributed by yz plane and the y and z Fourier transforms are performed without redistribution of data over processors.

transform proceeds as above then a global exchange of data is performed to arrange the data as shown in figure 2. The transformations in the y and z directions can then be performed without further redistribution of data. This scheme thus halves the communications requirement of the FFT compared with the first technique and is useful if the number of processors is relatively modest and the communications are slow compared with the compute speed of the processors.

It is well known that the performance of computer codes on vector machines is crucially dependent on the ratio of vector to scalar operations. The performance of codes on parallel machines is even more critically dependent on the ratio of parallel to sequential operations. Essentially this ratio must be greater than N:1 to benefit from the use of an N processor machine. It is clear that even the very smallest fraction of sequential operations will limit the size of machine on which calculations can be usefully performed. Figure 3 shows the operations involved in a total energy pseudopotential calculation performed using the conjugate gradients method. The letters in the boxes on the left of the figure show whether the operation is performed in real or Fourier space and the letters in the boxes on the right of the diagram show whether the operation is performed sequentially or in parallel. It can be seen that only a small number of set-up operations have to be performed sequentially. In particular the conjugate gradients loop shown in figure 3*b*, which dominates the computational cost of the total energy pseudopotential calculation, is executed totally in parallel. Unfortunately this is not always the case. In the case of metallic systems it is necessary to transform the trial wavefunctions to the Kohn–Sham eigenstates to determine the occupancies of the electronic bands. This transformation is determined using conventional matrix diagonalization routines which cannot be implemented efficiently on parallel machines. This technical problem has yet to be overcome.

The performance of total energy pseudopotential codes on the Daresbury Laboratory Intel iPSC/860 and the 'Grand Challenge' Meiko i860 Computing Surface at the University of Edinburgh are shown in figure 4. These timings are for small systems, containing just 64 silicon atoms in the unit cell, to allow comparison with the performance of a conventional supercomputer, one processor of a CRAY X-MP. Figure 4*a* is for the Meiko i860 computing Surface and figure 4*b* is for the Intel

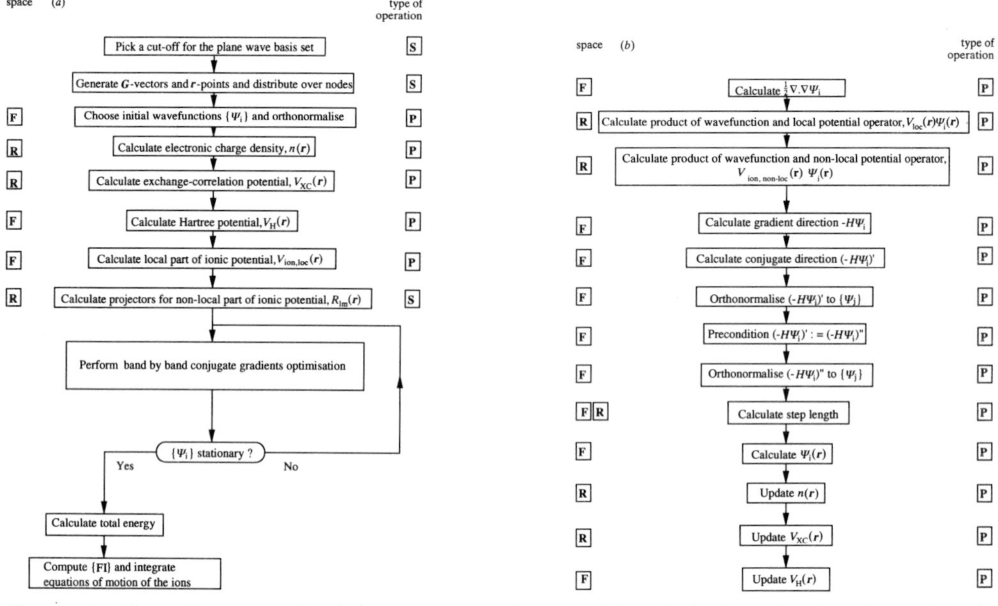

Figure 3. Flow diagram of total energy pseudopotential calculation showing (*a*) principle operations and (*b*) band by band conjugate gradients optimization. The letters in boxes on the left of the diagrams show whether the operation is performed in real space (R) or Fourier space (F) and the letters in boxes on the right of the diagram show whether the operation is performed in parallel (P) or sequentially (S).

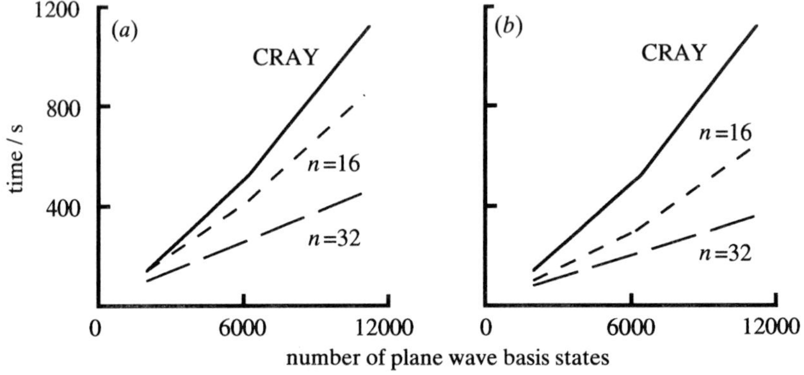

Figure 4. Variation of CPU time with number of plane wave basis states for total energy pseudopotential calculation for a 64 atom silicon cell. Figures show timings for 16 and 32 processors of (*a*) Meiko i860 Computing Surface and (*b*) Intel iPSC/860 against a single processor CRAY X-MP.

iPSC/860. The figures show the time required for one iteration of the conjugate gradients loop as a function of the number of plane wave basis states for $n = 16$ and $n = 32$ processors. It is clear that if the calculation is too small there is no advantage in using a larger number of compute nodes since the computational time is dominated by communications rather than compute speed. However, for larger calculations the advantage of using more processors is clearly shown in the figures. In particular, the Intel iPSC/860, which slightly outperforms the Meiko because the communications system is better suited to our particular application, 32 processors provide a performance three times faster than the CRAY X-MP.

4. The future

Already total energy pseudopotential calculations for 400 atom systems (in unit cells of 800 atomic volumes to allow for the presence of a surface) have been performed on a 64 node Meiko i860 Computing Surface (Stich *et al.* 1992) and a 16K Connection Machine (Brommer *et al.* 1992). These machines yield compute speeds on the total energy pseudopotential codes of the order of 1 GFlop. Intel have now installed the Delta machine at Caltech. This machine has 570 processors and a communication performance 10 times faster than that of the iPSC/860. This machine offers both compute and communications performance 10 times greater than the Meiko i860 Computing Surface used for the 400 atom calculations. Hence, the Delta machine would allow calculations to be performed for systems containing in excess of 1000 atoms. Intel have now announced the Paragon XP/S System which offers processing speeds of up to 300 GFlop and inter-processor communications speeds of 280 MBytes s^{-1}, a hundred times faster than the iPSC/860. This machine would be capable of carrying out calculations on systems containing several thousand atoms in the unit cell. With this machine many of the complex processes that require *ab initio* investigation will be within the power of the total energy pseudopotential method. Furthermore, these figures show that the teraFlop machine is an achievable goal.

This work was performed as part of the 'Grand Challenge' collaborative project, coordinated by Professor D. J. Tildesley. We acknowledge financial support from the Science and Engineering Research Council under grants GR/G32779, GR/G49524 and B18534:NACC. We are grateful for the help we have received from Dr N. M. Harrison and Dr W. H. Purvis at the Daresbury Laboratory, from Dr K. C. Bowler and Dr S. P. Booth at Edinburgh University, and from Mr D. Hewson at Meiko Scientific Ltd. We also thank Dr C. M. M. Nex for useful discussions.

References

Brommer, K., Needels, M., Larson, B. E. & Joannopoulos, J. D. 1992 *Phys. Rev. Lett.* **68**, 1355.

Car, R. & Parrinello, M. 1985 *Phys. Rev. Lett.* **55**, 2471.

Clarke, L. J., Stich, I. & Payne, M. C. 1992 *Comp. Phys. Commun.* (In the press.)

Cohen, M. L. 1984 *Phys. Rep.* **110**, 293.

Denteneer, P. & van Haeringen, W. 1985 *J. Phys.* C **18**, 4127.

Gillan, M. J. 1989 *J. Phys.* **1**, 689.

Heine, V. & Cohen, M. L. 1970 *Solid State Phys.* **24**.

Hohenberg, P. & Kohn, W. 1964 *Phys. Rev.* **136**, B864.

Ihm, J., Zunger, A. & Cohen, M. L. 1979 *J. Phys.* C **12**, 4409.

Joannopoulos, J. D. 1985 In *Physics of disordered materials*, p. 19. Plenum.

King-Smith, R. D., Payne, M. C. & Lin, J-S. 1991 *Phys. Rev.* B **44**, 13063.

Kohn, W. & Sham, L. J. 1965 *Phys. Rev.* **140**, A1133.

Payne, M. C., Teter, M. P., Allan, D. C., Arias, T. & Joannopoulos, J. D. 1992 *Rev. Mod. Phys.* (In the press.)

Phillips, J. C. 1958 *Phys. Rev.* **112**, 685.

Pickett, W. 1989 *Comp. Phys. Rep.* **9**, 115.

Stich, I., Payne, M. C., King-Smith, R. D., Lin, J-S. & Clarke, L. J. 1992 *Phys. Rev. Lett.* **68**, 1351.

Teter, M. P., Payne, M. C. & Allan, D. C. 1989 *Phys. Rev.* B **40**, 12255.
Wilkinson, J. H. 1965 *The algebraic eigenvalue problem*. Oxford: Clarendon.

Discussion

A. M. Stoneham (*Harwell Laboratory, Didcot, U.K.*): In addition to your approach to increasing the number of atoms or electrons considered, there are other routes to the really useful level, namely embedding (of which Green's function methods are a special case) and mesoscopic modelling (Harding, this volume).

M. C. Payne: We agree that the two approaches suggested are useful. Even with thousands of atoms in a unit cell *ab initio* calculations will still be limited to the nanometre lengthscale and alternative methods must be used to study longer lengthscale systems. Mesoscopic modelling attempts to study such systems by parametrizing the behaviour of the system on these longer lengthscales. The values of the parameters should be determined by independent experiment (rather than adjusting the parameters so that the model 'agrees' with experiment) but in many cases this is not feasible and then the results of the modelling may be questionable. *Ab initio* calculations provide an alternative method for determining the values of physical parameters that cannot be determined experimentally and so combining *ab initio* calculations with mesoscopic or macroscopic modelling provides a technique for extending *ab initio* studies to large systems and makes such modelling significantly more realistic and rigorous. Embedding schemes provide an efficient method for studying isolated defects in bulk systems. However, present embedding schemes only allow ionic relaxation within the embedding surface as this surface separates the defective region from the perfect bulk. In favourable cases, such as some close-packed metal surfaces where ionic relaxation is essentially limited to one or two atomic layers, only a few atoms are required within the embedding surface. Less favourable situations might require as many as a thousand atoms within the embedding surface and considerable work is required before enbedding techniques can be applied to systems containing this number of atoms.

Crystal excitation: survey of many-electron Hartree–Fock calculations of self-trapped excitons in insulating crystals

By A. L. Shluger[1,2], A. H. Harker[3], R. W. Grimes[2] and C. R. A. Catlow[2]

[1] *Department of Chemical Physics of Condensed Matter, University of Latvia, 19 Rainis blvd, Riga, Latvia*

[2] *The Royal Institution of Great Britain, 21 Albemarle Street, London W1X 4BS, U.K.*

[3] *AEA Industrial Technology, Harwell Laboratory B424.4, Didcot, Oxfordshire OX11 0RA, U.K.*

To model successfully the diversity of electronic structure exhibited by excitons in alkali halides and in oxide materials, it is necessary to use a variety or combination of theoretical methods. In this review we restrict our discussion to the results of embedded quantum cluster calculations. By considering the results of such studies, it is possible to recognize the general similarities and differences in detail between the various exciton models in these materials.

1. Introduction

The elementary excitations of some insulating crystals form an interesting bridge between those of molecules, which are necessarily confined by the size of the molecule, and those of semiconductors, in which the electrons, holes and excitons (bound electron-hole pairs) are free to travel through the lattice. In insulating crystals such as alkali-metal halides, alkali-earth fluorides, silica and others, the excitons and holes, or only excitons, may be self-trapped to produce small radius polaron states. This self-trapping into a state which resembles an atom or molecule is accompanied by strong local lattice relaxation, and drastically alters the optical, luminescent and energy transport properties of the crystal.

Some quite basic questions concerning the self-trapped exciton (STE) remain unresolved, even in the alkali halides which are traditionally regarded as prototype insulating materials in which the microscopic features of the self-trapping processes have been studied most extensively. It is known that a self-trapped exciton in insulating crystals is created after the relaxation of a free exciton or the trapping of an electron by a self-trapped hole. Elaborate experimental studies on the STEs have been carried out during the last few decades, using optical absorption, luminescence, ENDOR, dynamics of formation in picosecond and femtosecond ranges, resonant Raman scattering and other techniques, as reviewed by Williams & Song (1990) and Itoh & Tanimura (1990). Nevertheless experimental determination of the model of the STE has not been successful so far. Therefore the model of the STE, i.e. its atomic and electronic structure, is still the main goal of theory.

The general criteria for the localization/delocalization of excitons are developed on the basis of the phenomenological theory of excitons in phonon fields, using the

Phil. Trans. R. Soc. Lond. A (1992) **341**, 221–231

Printed in Great Britain
© 1992 The Royal Society

Frohlich hamiltonian and more sophisticated methods, as reviewed by Ueta *et al.* (1986). The structure and properties of the self-trapped species, however, may be studied using a static quantum mechanical approach, and it is such calculations which are reviewed in this paper.

2. Calculation techniques

(*a*) *Basic ideas*

The very possibility of this approach is based on the assumption that the exciton is well localized in a small area of the crystal. However, even if it is true, there are apparent distinctions between the ground or excited state of the localized point defect and the STE.

Self-trapped excitons are the short-living luminescent states of excited crystals, which have properties contrasting with those for the crystal ground state. Approximately, the initial states for the luminescence may be treated as stationary and correspond to the minima of the adiabatic potential of the excited crystal. The atomic structure of the excited crystal at these minima are often considered as models for STEs. (We shall not discuss the quenching of the exciton luminescence, although it also provides an insight into the STE structure as was shown by Shluger *et al.* (1991*c*).)

In the alkali-metal halides both singlet and triplet states of the self-trapped exciton have been detected experimentally. In crystalline quartz only the triplet state of the STE has been observed so far, whereas in MgO the multipole state of the exciton is not yet clear. Nevertheless most theoretical studies concern only the triplet excited state of the crystal. This is solely for technical reasons. In wide-gap insulators, as indeed all of these crystals are, the crystal ground triplet state automatically means the excited state. For the lowest multiplicity it may be treated in the single-determinant unrestricted Hartree–Fock approximation. This idea has been used in all the many-electron calculations to date. Conversely, the singlet excited state must be treated using a many-determinant approximation for the wave-function of the crystal. This is much more time-consuming and the first calculations for the singlet state of the STE in several alkali-metal halides have been made only recently using the generalized valence bond (GVB) technique by Song & Baetzold (1992).

The self-trapping of the exciton in many cases is accompanied by its decomposition. The electron and the hole, overlap strongly, but become localized in the lattice for different reasons. According to electron spin resonance (ESR) experimental data, the electron of the STE is much more delocalised than the hole. The former is localized by the crystalline potential and the degree of its localization depends sharply on the relative position of the hole. Special effort is needed in order to take this into account.

(*b*) *Background*

As in the case of experimental studies, most attention has in the past been paid to the STE in alkali halides. The defect here was believed to have D_{2h} symmetry, and the atomic structure of an electron trapped by the Coulomb field of the positively charged self-trapped hole, i.e. a V_k centre (an X_2^- molecular ion occupying two anion sites, where X denotes a halogen atom). This is the (V_k+e) model.

Historically some of the first calculations of the electronic structure of the STE in this model were performed using the many-electron Hartree–Fock method. The idea of the calculations performed by Wood (1966) was to understand the key experimental fact that the luminescence of the STE has a large Stokes shift compared with the exciton excitation energy. These calculations contained all the basic ideas which were used in most further studies: (i) the exciton was treated within the same calculation technique as was developed for the calculations of point defects in insulating crystals; it also accounted for the interaction of the exciton with the polarization of the remaining crystal. (ii) The comparison of the two possible models for the qualitative interpretation of the key experimental facts. It was indeed the first simulation of the STE, which provided strong support for the $(V_k+\mathrm{e})$ model. A similar approach has been applied recently for the studies of the STE models in MgO and SiO_2.

The Hartree–Fock calculations by Stoneham (1974) were performed for a molecular cluster Na_2Cl_2 and treated both singlet and triplet states of the $(V_k+\mathrm{e})$ model of the STE in NaCl. Their aim was to produce results which could be used as a test for the simpler and less sophisticated approaches that will be necessary in more complex situations. This strategy has proved to be very successful and was used in further studies of STE models.

In particular, Song and Leung developed the one-electron approach, in which the STE was treated in an extended ion approximation using the hybrid-potential method. The specific feature of this approach was that it emphasized the role of the electron component of the STE and treated the hole as a frozen quasi-molecule. It has been shown that the X_2^- molecular ion of the STE is displaced along the (110) axis in a wide class of alkali halides and thus the STE has a C_{2v} symmetry and its structure is more like a nearest-neighbour pair: an F centre and an H centre (i.e. an X_2^- molecular ion occupying a single anion site) rather than $(V_k+\mathrm{e})$. The displacement of the X_2^- molecular ion from the $(V_k+\mathrm{e})$ configuration arises from the repulsive character of the non-local pseudo-potential of the X_2^- molecular ion for the STE electron and the attraction of the electron to the anion vacancy formed after the off-centre displacement of the X_2^- molecular ion. A similar structure is characteristic of the STE in alkali-earth halide crystals. A thorough review of this approach has been presented in recent publications by Song (1991) and Williams & Song (1990).

The theory developed by Song *et al.* has given a reasonable account of a large number of the experimental data. It has also shed light on several important questions which remained unsolved. In particular, in the framework of the one-electron approach the interaction between the two unpaired electrons of the STE is not treated explicitly. The exchange and correlation between these two electrons strongly depend on the electronic structure of the hole component of the STE and plays a major role in the process of exciton self-trapping and its decay into the Frenkel pair of the primary defects. Both effects may be treated only by many-electron theory.

(c) *Method of calculation*

Two basic models of the defective lattice were used in the many-electron studies of excitons: the molecular cluster model and the periodic defect model. To present a brief account of the computation techniques we need to give a few details of the methods and computer codes.

ICECAP and CLUSTER codes combine the unrestricted Hartree–Fock (UHF) method of calculating the electronic structure of the molecular cluster with the classical Mott–Littleton approach which accounts for the polarization of the remaining crystal (see Vail 1989; Vail *et al.* 1991 for review). One of the basic ideas of the Mott–Littleton approach is to divide the crystal into two regions: an inner region (I), containing the defect and its immediate surround, and an outer region (II) which responds as a dielectric continuum. The molecular cluster is placed at the centre of region I. There are the following basic differences between the two codes: (i) the ICECAP code uses the *ab initio* UHF method developed by Kunz & Vail (1988), whereas the CLUSTER code is based on the semi-empirical version of the UHF method implementing the approximation of intermediate neglect of differential overlap (INDO) as described by Shluger & Stefanovich (1990); (ii) the lattice ions in region I outside the cluster are treated in the ICECAP code within the shell-model approximation, as described by Vail *et al.* (1984), and in the CLUSTER code using the non-point polarizable ion approximation developed by Kantorovich *et al.* (1988). The ICECAP code was used for the calculation of the STE in alkali-metal halides, whereas the CLUSTER code was used for STE studies in oxides SiO_2, MgO and Li_2O.

The DICAP code uses the same ideas as the ICECAP and CLUSTER codes (see Puchin *et al.* 1992), but does not account for lattice polarization outside the cluster. Therefore it was extensively used to search for the extreme points on the APES in NaCl, and then further calculations using the ICECAP code were made near the extrema for comparison.

In both ICECAP and DICAP codes cations and anions inside the cluster are represented by the semi-local norm-conserving pseudopotentials of Bachelet *et al.* (1982). The same representation was used for the cations outside the cluster; this is essential to prevent unphysical delocalization of the wave-function of the excited electron. Anions may be treated either by the whole-ion pseudopotential approximation or considered as point ions.

A similar but much more simplified approach is used in the *ab initio* CAD-PAC code, which was applied recently to the calculation of the STE in several alkali halides by Song & Baetzold (1992). In this code the quantum-mechanical cluster is embedded within an array of fixed point ions. To prevent the surface atoms of the quantum cluster from undergoing unphysical displacements in the course of geometry optimisation, Born–Mayer repulsive potentials are included between all of the fixed ions adjacent to and interacting with the quantum-mechanical ions.

To follow the changes in the electron distribution, the basis set has to be flexible enough to allow different types of localization. As has been first suggested by Baetzold & Song (1991) this may be achieved by using a set of floating gaussian orbitals (FGOS) centred in different points within the cluster. This method was used in all *ab initio* calculations.

The CLUSTER code combines the possibilities of carrying out calculations of point defects within the embedded cluster and periodic models. The periodic model has an advantage of preserving the equivalence of the perfect lattice. The latter is essential for the simulation of exciton self-trapping. The main limitation of the model is the mutual perturbation of periodically arranged defects. This can be overcome by considering large periodic cells and within simplified semi-empirical techniques. This method has been applied to the study of the STE and the nearest defect pair in SiO_2 by Shluger & Stefanovich (1990).

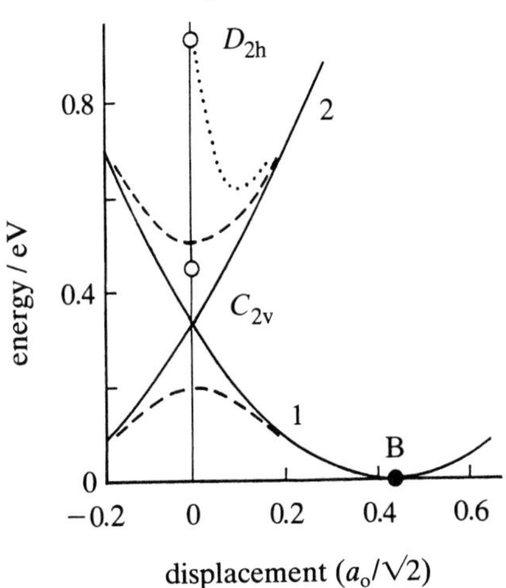

Figure 1. Section of the APES for the lowest triplet state of the STE in NaCl. The energy curves are presented in the coordinates: the displacement of the centre of mass of the Cl_2^- molecular ion along the (110) crystalline axis against the change of the total energy relative to the STE minimum (point B). All other configuration coordinates, which determine the total energy of the system, including the molecular bond length, the displacements of the surrounding ions, the positions and exponents of FGOs (see figure 2) are optimised at each point of the APES. 2 corresponds to the UHF solution with the opposite polarisation of the electron and the hole of the STE to that in 1. Dotted line corresponds to the compact model of the STE in the vicinity of the D_{2h} configuration (see Shluger *et al.* 1991; Puchin *et al.* 1992 for discussion). The D_{2h}, C_{2v} and crossing point at zero displacement of the centre of mass of the Cl_2^- molecule correspond to solutions for which the nuclei and electron wavefunctions have $[D_{2h}, D_{2h}]$, $[D_{2h}, C_{2v}]$ and $[C_{2v}, C_{2v}]$ symmetries respectively.

3. Properties of self-trapped excitons

(*a*) *Atomic and electronic structure*

(i) *Alkali-metal halides*

Since it has been generally assumed that the STEs in alkali halides consist of a diffuse electron bound to a V_k centre, this model was used as a starting point in all theoretical studies. The results of recent *ab initio* calculations of the lowest triplet state of the (V_k+e) model in several alkali halide crystals (LiCl, NaF, NaCl, NaBr, KCl) performed by Baetzold & Song (1991), Shluger *et al.* (1991*b*), Song & Baetzold (1992) and Puchin *et al.* (1992) have indicated that in the framework of the UHF method this configuration is unstable with respect to the displacement of the V_k centre along the (110) crystalline axis from its original position, which had D_{2h} symmetry. However, as has been discussed by Puchin *et al.* (1992), the single-determinant UHF method fails in the vicinity of the point in the configurational space of the (V_k+e) system where the nuclei have D_{2h} symmetry (henceforth we will call it a D_{2h} point). In fact the interaction of several electronic configurations must be taken into account at this point to obtain a correct electronic energy. Nevertheless, we will assume that electron correlation is not vital far from this point and will make a critical analysis of the results obtained for the strongly off-centred configuration of the STE and the F–H pair. Some of these results will allow us to make additional comments regarding the D_{2h} point.

The most complete set of experimental data exists for the STE in NaCl. Therefore it was studied in greatest detail. The section of the APES for the lowest triplet state of NaCl, calculated by Puchin *et al.* (1992) in a wide range of configuration

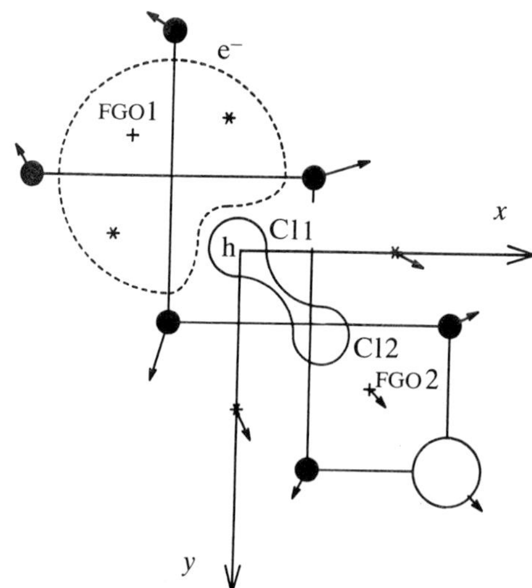

Figure 2. Schematic presentation of the off-centre model of the STE in alkali halide crystals. Arrows show the directions of displacements of ions from their lattice sites, and of floating orbitals from their positions in the D_{2h} configuration. The floating orbitals FGO1 and FGO2 are located in the (110) plane. ○, anions; ●, cations; ✱, pairs of FGOs with coordinates $(x, y, \pm z)$.

Table 1. *The calculated properties of the STE in several alkali halide crystals*

	structural properties and the frequencies of stretching vibration of the hole component of STE				optical excitation energy/eV		triplet luminescence energy/eV	
crystal	$R_{\mathrm{C11\text{-}C12}}$/nm	ΔQ_2/nm	δ/nm	ω_{str}/cm^{-1}	theor.	exp.[c]	theor.	exp.[c]
LiCl[a]	0.265	0.115	0.163	—	2.8	2.2	4.4	4.18
NaF[b]	—	0.198	—	—	—	—	2.4	2.7
NaCl[a]	0.264	0.087	0.153	218	2.6	1.95	3.8	3.51
NaCl[b]	—	0.099	—	—	—	—	3.8	—
NaBr[b]	—	*ca.* 0.12	—	—	—	—	2.9	4.6
KCl[a]	0.277	0.126	0.208	206	2.0	1.87	3.2	2.31

[a] Results of Shluger *et al.* (1991*b*) and Puchin *et al.* (1992).
[b] Results of Baetzold & Song (1992).
[c] Results of Williams & Song (1990).

coordinates using the DICAP code, is shown in figure 1. The atomic configuration corresponding to the minimum B of the APES is schematically depicted in figure 2. The analysis of the APES indicates that both translational motion and rotation of the hole component of the STE are important for the determination of the positions of extreme points of the adiabatic potential. For LiCl, KCl, NaF and NaBr only the part of the APES near the minimum has been studied; this is usually attributed to the STE (point B in figure 1). Let us first discuss this part of the APES.

Some of the numerical data, characterizing the minima of the APES, are summarized in table 1, which is to be considered together with figure 2. These results were obtained by the DICAP, ICECAP and CADPAC methods. All three methods gave similar results. The electron component of the exciton is localized mainly in the anion vacancy, although about 15% of the spin density is contributed by the s orbitals of the cations surrounding the anion vacancy. More than 80% of the spin density of the hole is localized on Cl1. Thus these minima represent qualitatively the model of the triplet excited state of the crystal, corresponding to preferential localization of the hole on one anion of the X_2^- molecular ion, occupying an intermediate position between the V_k and H-centre configurations (ΔQ_2 corresponds to the displacement of the centre of mass of the Cl_2^- molecule). The excited electron, on the other hand, is attracted to the vacant anion site adjacent to the hole.

The distance between chlorine ions, Cl1 and Cl2 (see figure 2), in the Cl_2^- molecular ion, is slightly different from that in the V_k-centre, but still much larger than in the H-centre. Strong polarization makes the chemical bond in Cl_2^- weaker which results in smaller frequencies of the stretching vibrations as compared with the V_k-centre. We should note that the distances, δ, between the vacant site and the anion Cl1 in LiCl and NaCl are similar and substantially smaller than in KCl (see table 1). The calculated energies of Franck–Condon transitions, corresponding to the π polarized lowest optical excitations of the electron component and triplet luminescence (vertical transition into the ground singlet state), are presented in the last two columns of table 1.

Curve 2 in figure 1 corresponds to the configuration of the STE in which the hole is localized on Cl2 and the electron is near the anion vacancy adjacent to the hole. The points on curves 1 and 2 that have equal total energies represent two physically equivalent configurations of the electron and the hole. The single-determinantal wave-functions of these configurations are not orthogonal. This causes an artificial crossing of the two curves. In the crossing region the substantial interaction between these configurations (or in other words the electron correlation) should lead to a splitting of the two curves (in fact surfaces), which is qualitatively depicted in figure 1 by two broken curves. These may be obtained using the configuration interaction technique developed for non-orthogonal determinantal wave-functions by Broer *et al.* (1991). Therefore the shape of the APES as well as the validity of the adiabatic approximation in this region requires further study.

Since the D_{2h} state of the triplet exciton is apparently unstable, these comments are relevant first of all to the mechanism of diffusion of the STE. Within the many-electron approach this was studied only for the STE in NaCl. As has been pointed out by Chen & Song (1990), in the framework of the off-centre model of the STE, the mechanism for its diffusion in alkali halides may be considered as being comprised of two steps: (i) a joint motion of the electron and the hole along the (110) axis (the barrier for this was just discussed in the previous paragraphs); and (ii) a rotation of the hole component of the STE which changes its direction by 60°, which very much resembles the 60° reorientation of the V_k-centre. The simulation of the diffusion mechanism of the STE in NaCl was performed using the DICAP code. The calculated height of the barrier for the 60° reorientation of the hole component of the STE was equal to 0.44 eV. This is much larger than the activation energy for diffusion of the STE in NaCl (0.13 eV) (observed experimentally by Tanimura & Itoh (1981)) but close to the characteristic values for V_k-centre migration (0.54 eV for KCl). These results show clearly that the exciton relaxation energy and even the shape of its APES can not yet be calculated as accurately as necessary in all cases.

(ii) *Oxides*

Let us now compare the STE model in alkali halide crystals with that in SiO_2 which has a different electronic and crystalline structure. The self-trapping of triplet excitons in crystalline and amorphous SiO_2 has been well established experimentally.

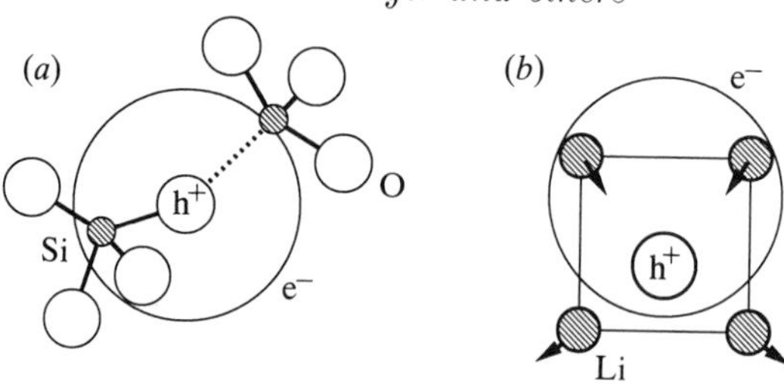

Figure 3. Schematic models for exciton structures in (a) SiO_2, (b) Li_2O.

The results of recent theoretical calculations by Fisher (1989), Shluger (1988) and Shluger & Stefanovich (1990*b*) are consistent with exciton self-trapping being accompanied by the weakening of one of the two Si–O bonds of an oxygen, in addition to the displacement of the oxygen ion (by 0.03 nm) towards an interstitial position. The electronic and spin density distributions at the minimum of the APES, described in these calculations correspond to almost complete localization of the hole component of the STE on the displaced oxygen. The rather diffuse electron component of the STE is centred at the anion vacancy (see figure 3*a*). The difference in total energy of the ground (singlet) state and the triplet state at the configuration corresponding to the energy minimum of the APES amounts to 2.7 eV. This is the Frank–Condon transition energy from the triplet to the singlet term of the system which compares favourably with the peak position (about 2.8 eV) of the exciton luminescence (see Williams & Song 1990 for a review).

A similar model to that described for SiO_2 has been predicted by Shluger & Itoh (1990) for the STE in Li_2O which has the antifluorite structure and ionic bonding. In this case, exciton self-trapping is accompanied by an almost complete localization of the hole to the oxygen which is displaced slightly along the ⟨110⟩ axis towards the interstitial position. The electron is localized around the lattice site of this displaced oxygen (see figure 3*b*).

In SiO_2 and Li_2O the large lattice relaxation, resulting from the unscreened hole leads to a large (often more than 3 eV) Stokes shift of the excitonic luminescence. The model often used to explain the experimentally observed excitonic-type excitation spectra for MgO is the on-centre oxygen 2p–3s excitation (see, for example, Vail *et al.* 1991). To determine whether the on-centre model of STE can also best explain the experimentally observed relatively small Stokes shift of luminescence (about 0.75 eV), several models of the triplet excitation of MgO were simulated by Shluger *et al.* (1991*a*) using the CLUSTER code. These calculations have shown that the on-centre model of the exciton, in which the hole is screened by the electron, is able to offer a reasonable fit to the available experimental data. Good agreement was found between the calculated (0.65 eV) and experimental Stokes shifts and also between the calculated exciton luminescence energy, 6.95 eV, and the experimental value, 6.9 eV.

(*b*) *Defect formation and recombination*

The creation of primary Frenkel defects, and their separation and recombination constitutes a broad class of problems, which have been studied rather extensively. The main emphasis has been on the kinetic side of the problem, and only a few microscopic calculations have been made so far. The idea of the calculations by Itoh *et al.* (1977, 1980) was to study the role of the electron and hole excited states in the STE decomposition in alkali halides. It is known that F and H centres in alkali halides are formed not only from the ground state of the STE, but also from some of the

TRINITY SCIENCE LIBRARY

excited states. Itoh & Saidoh have suggested that there is an almost barrierless downhill potential leading from the hole excited state to the F and H centre pair.

The calculations by Itoh *et al.* (1977, 1980) were performed to check this possibility. Although the significance of the excited states of the STE in the defect formation is still not clear, some of the ideas suggested in these papers are nevertheless important. They used the CNDO semi-empirical calculation scheme and suggested a method of parametrization based on the crystal and defect properties. For the F centre calculation, the basis set of atomic orbitals was extended by a Slater-type orbital centred at the anion vacancy. This idea proved to be very fruitful and has subsequently been used in many other studies. Transfer of excitation energy from the electron to the hole components of the STE was studied quantitatively on the basis of the quantum-chemical calculations of the transition matrix elements.

The possibility of radiative tunnelling transitions (RTT) between close F and H centres was examined by Shluger *et al.* (1992). In these calculations an earlier version of the CLUSTER code was used. The luminescence due to the RTT of an electron between well-separated defects is a well-known effect in both semiconductors and ionic crystals. Its peculiarity is that the RTT occurs directly between ground electronic states of the donor and acceptor as a result of which the emitted energy depends on properties of both partners. The calculations by Shluger *et al.* (1982) predicted a small luminescence energy for the RTT between F and H centres in LiF and KCl. They emphasized the difference in the polarization of the lattice by the F–H pair and the α-I pairs of defects.

A detailed *ab initio* study of the APES for further separation of the electron and the hole components of the STE to form the nearest neighbour and the more distant F–H pair was performed only for NaCl and KCl by Puchin *et al.* (1992). In the KCl case only the adiabatic diffusion of the hole component along the (110) crystal axis was considered. The principal results of this study, which were made using the ICECAP code, may be summarized as follows: (i) the next minimum on the APES along the (110) axis is located at the position corresponding to the next nearest-neighbour F–H pair; (ii) the barrier that the hole needs to overcome to reach this minimum, diffusing straight along the axis, was calculated to be 0.54 eV, which seems too large; (iii) at the next nearest-neighbour configuration the ground singlet state of the crystal, calculated with account of lattice polarization, is located about 0.1 eV lower than the triplet state. This excludes a luminescence from this state.

A more comprehensive study has been made for NaCl, using the DICAP code. These results clearly emphasize the importance of the reorientation of the H-centre. The most compact stable configuration of the nearest F and H centres corresponds to the H-centre located in the site nearest to the F-centre anion and reoriented by 90° with respect to the initial STE orientation along the (110) axis. The H-centre in this configuration is oriented along the (111) axis. The luminescence energy, corresponding to the electron-hole recombination was calculated to be close to zero.

4. Problems and discussion

With respect to the accuracy of the theoretical methods, the single-determinant UHF method and the pseudopotential technique, have made an extensive study of the APES for the STE and the nearest defect pairs possible. However, as is clear from the results, electron correlation and the polarization of the cores of cations (especially for KCl) may play an important role. However, some qualitative conclusions may be made on the basis of the results obtained in the reviewed studies. (i) The reorientation of the hole component of the STE is an important factor which must be taken into account in the construction of the atomistic model of the STE in alkali halides. (ii) The electronic states of the STE which are responsible for the luminescence of the exciton in alkali halides are located within the limits of the configuration coordinates corresponding to the $(V_k + e)$ and the nearest-neighbour F–H pair. (iii) The off-centre model of the STE in alkali halides is valid but the real position of the hole component of the STE and the degree of its polarization cannot be reliably evaluated by the present technique. Despite this, the reasonable agreement between the calculated optical excitation and luminescence energies with the experimental data suggest that the models for the STE in LiCl, NaCl and KCl may not change drastically after the method is improved. The consistency of the results for five crystals also suggests that the present models may provide a good basis for further discussion of the experimental facts. (iv) Self-trapped excitons in wide band insulators are composed of localized electron-hole pairs, whose properties have certain general features. The hole state is mostly localized on one anion displaced from its lattice site. The exciton electron component is localized around the vacant anion site which is adjacent to the hole. The extent of the ion displacements, which can be quite different, depend upon the system in question and influence the degree to which the electron component is localized.

Finally, we should note that the accuracy of the many-electron calculations at the present stage is not high enough to describe fully the electronic and atomic structures of the relaxed configurations of the STEs. It is very important to make clear the elementary processes that play a role in the dynamic properties of the STEs. For instance, the pseudo Jahn–Teller instability of the one-centre and the $(V_k + e)$ models of the STE are not adequately understood in the circumstance where the polarization of the hole plays a role. The most fruitful approach might combine both the many-electron treatment with model hamiltonian methods.

This work is supported partly by Grant-in-Aid for Inter-University Collaboration Program of Ministry of Education, Science and Culture of Japan. A. L. S. acknowledges the Canon Foundation in Europe and the Royal Society for financial support. The work of A. H. H. forms part of the underlying programme of Corporate Research of AEA Technology. We are grateful to V. E. Puchin, N. Itoh, R. Baetzold and A. M. Stoneham for valuable discussions.

References

Chen, L. F. & Song, K. S. 1990 A theory of hopping diffusion of the self-trapped exciton in alkali halide crystals. *J. Phys.* A **2**, 3507–3513.

Bachelet, G. R., Hamann, D. E. & Schluter, M. 1982 The pseudopotentials that work: H to Pt. *Phys. Rev.* B **26**, 4199–4211.

Baetzold, R. C. & Song, K. S. 1991 A study of the structure of self-trapped exciton in alkali halides by *ab initio* method. *J. Phys.* A **3**, 2499–2505.

Broer, R., Van Oosten, A. B. & Nieuwpoort, W. C. 1991 Nonorthogonal CI description of localized excitations in ionic transition metal compounds. *Rev. Solid State Sci.* **5**, 79–90.

Fisher, A. J. 1989 Theoretical studies of point defects. *AERE report*, R13770, ch. 6.

Itoh, N., Stoneham, A. M. & Harker, A. H. 1977 The initial production of defects in alkali halides: F and H centre production by non-radiative decay of the self-trapped exciton. *J. Phys.* C **10**, 4197–4209.

Itoh, N., Stoneham, A. M. & Harker, A. H. 1980 Non-radiative de-excitation of highly excited self-trapped excitons in alkali halides: Mechanism of the F and H centre production. *J. Phys. Soc. Japan* **49**, 1364–1371.

Itoh, N. & Tanimura, K. 1990 Formation of interstitial-vacancy pairs by electronic excitation in pure ionic crystals. *J. Phys. Chem. Solids* **51**, 717–735.

Kantorovich, L. N. 1988 An embedded-molecular-cluster method for calculating the electronic structure of point defects in non-metallic crystals: I. General theory. *J. Phys.* C **21**, 5041–5056.

Kunz, A. B. & Vail, J. M. 1988 Quantum-mechanical clustet-lattice interaction in crystal simulation: Hartree–Fock method. *Phys. Rev.* B **38**, 1058–1063.

Puchin, V., Shluger, A., Tanimura, K. & Itoh, N. 1992 On the model of the self-trapped exciton in alkali halides. *Phys. Rev.* B (Submitted.)

Shluger, A., Kotomin, E. & Kantorovich, L. 1982 Calculation of energies of radiative tunneling transitions between defects in alkali halides. *Solid State Commun.* **42**, 749–752.

Shluger, A. 1988 The model of a triplet self-trapped exciton in crystalline SiO_2. *J. Phys.* C **21**, L431–L434.

Shluger, A. & Itoh, N. 1990 Models of electronic defects and self-trapped excitons in Li_2O. *J. Phys. Condensed Matter* **2**, 4119–4125.

Shluger, A. & Stefanovich, E. 1990 Models of the self-trapped exciton and nearest-neighbor defect pair on SiO_2. *Phys. Rev.* B **42**, 9664–9673.

Shluger, A., Grimes, R. W., Catlow, C. R. A. & Itoh, N. 1991*a* Self-trapping holes and excitons in the bulk and on the (100) surfaces of MgO. *J. Phys. Condensed Matter* **3**, 8027–8036.

Shluger, A., Itoh, N., Puchin, V. & Heifets, E. 1991*b* Two types of self-trapped excitons in alkali halide crystals. *Phys. Rev.* B **44**, 1499–1508.

Shluger, A., Georgiev, M. & Itoh, N. 1991*c* Self-trapped excitons and interstitial-vacancy pairs in oxides. *Phil. Mag.* **63**, 955–964.

Song, K. S. 1991 Study of excited electronic states in insulators. *Rev. Solid State Sci.* **5**, 477–494.

Song, K. S. & Baetzold, R. C. 1992 Structure of the self-trapped exciton and nascent Frenkel pair in alkali halides: an *ab initio* study. *Phys. Rev.* B **46**, 1960–1969.

Stoneham, A. M. 1974 Electronic structure of the self-trapped exciton in sodium chloride. *J. Phys.* C **7**, 2476–2486.

Tanimura, K. & Itoh, N. 1981 The hopping motion of the self-trapped exciton in NaCl. *J. Phys. Chem. Solids* **42**, 901–910.

Tanimura, K. & Itoh, N. 1984 Selective non-radiative transitions at excited states of the self-trapped exciton in alkali halides. *J. Phys. Chem. Solids* **45**, 323–340.

Ueta, M., Kanzaki, H., Kobayashi, K., Toyozawa, Y. & Hanamura, E. 1986 *Excitonic processes in solids*. Berlin: Springer-Verlag.

Vail, J. M., Harker, A. H., Harding, J. H. & Saul, P. 1984 Calculations for electronic point defects with self-consistent lattice polarization. *J. Phys.* C **17**, 3401–3414.

Vail, J. M. 1989 Theory of electronic defects: Applications to MgO and alkali halides. *J. Phys. Chem. Solids* **51**, 589–607.

Vail, J. M., Pandey, R. & Kunz, A. B. 1991 Embedded quantum cluster simulation of point defects and electronic band structures in ionic crystals. *Rev. Solid State Sci.* **5**, 241–283.

Williams, R. T. 1989 Creation of lattice defects in the bulk and on the surface of alkali halide crystals. *Radiat. Effects Defects Solids* **9**, 175–187.

Williams, R. T. & Song, K. S. 1990 The self-trapped exciton *J. Phys. Chem. Solids* **51**, 679–716.

Wood, R. F. 1966 Luminescence from exciton and V_k-plus-electron states in alkali halide crystals. *Phys. Rev.* **151**, 629–641.

Direct free energy minimization methods: application to grain boundaries

By A. P. Sutton
Department of Materials, Oxford University, Parks Road, Oxford OX1 3PH, U.K.

A critical review is given of recently developed methods for determining the atomic structures and solute concentration profiles at defects in elemental solids and substitutional alloys as a function of temperature. Exact results are given for the effective force on an atom arising from the vibrational entropy in the quasiharmonic approximation and for the occupancy of a site in the pair potential approximation. An improved, approximate formula is given for the effective force arising from the vibrational entropy. The mean field approximation that is used in the alloy problem is compared with the auto-correlation approximation. It is shown that the better statistical averaging of the auto-correlation approximation leads to effective pair interactions that are temperature and concentration dependent.

1. Introduction

Kinetic, mechanical and electrical properties of grain boundaries in most (possibly all) materials vary markedly with temperature and purity of the specimen. In elemental bicrystals the excess vibrational entropy of an interface can drive grain boundary phase transformations. The introduction of minute quantities of impurities can result in strong grain boundary segregation and dramatic changes in grain boundary properties. The modelling of such processes entails the consideration of the appropriate free energy and its minimization with respect to the atomic coordinates and the local impurity concentration. This is a formidable task because of the *five dimensional* parameter space that characterizes the orientation of the boundary plane and the misorientation between the crystal lattices. Faced with such a task it is essential to focus on *trends in behaviour* and to avoid excessively detailed studies of a few rather special cases (for an example of such an approach see Sutton (1991)). In this paper I review some new methods that involve direct minimization of free energy functionals (Sutton 1989; LeSar *et al.* 1989; Najafabadi *et al.* 1991*a*, *b*). The much greater computational speed of these methods, compared with other available techniques (see Rickman & Phillpot 1991, and references therein), and the relative ease with which all excess thermodynamic state variables are obtained, offer the greatest feasibility for exploring trends in behaviour. My purpose is to give a more thorough discussion than has so far appeared of the approximations and physical assumptions involved in these new methods.

The essence of the 'direct methods' is (i) to write down a functional form for the free energy of the system, in terms of microscopic variables such as the average atomic coordinates and atomic site occupancies, at a given temperature, pressure and set of chemical potentials, and (ii) to minimize this functional form by using the derivatives of the free energy with respect to the microscopic variables. The

Phil Trans. R. Soc. Lond. A (1992) **341**, 233–245 © 1992 The Royal Society computational effort is little more than that involved in a minimization of the energy of a system at 0 K, and standard steepest descent, conjugate gradient or variable metric energy minimization techniques may be used.

2. Direct free energy minimization in an elemental solid

Here the issue is the determination of the atomic structure and excess thermodynamic quantities of a grain boundary (or some other defect) in an elemental solid at an elevated temperature. Since atoms are vibrating we have to define what we mean by the atomic structure. This is true even at 0 K owing to the zero point energy. We mean the time averaged positions of the atoms, assuming that the system is in thermodynamic equilibrium, and assuming that diffusion does not take place. The time averaged structure is what is measured in an X-ray diffraction experiment for example. As the temperature changes the time averaged positions of the atoms change until the time averaged forces on them are zero. The time averaged forces vary with temperature because of the anharmonicity of the atomic interactions. The time averaged structure may be determined, therefore, by requiring that the time averaged force on each atom is zero. By ergodicity the time averaged value of a quantity is identical to the ensemble average. The ensemble average of the force on an atom is equal to the negative of the gradient of the ensemble free energy with respect to its position. Thus, by expressing the free energy of the system as a function of the average atomic coordinates, we can obtain the time averaged structure directly by simply minimizing the free energy with respect to the position of each atom (Sutton 1989; LeSar *et al.* 1989).

It is emphasized that it is assumed that each atom never leaves its own potential well. More precisely, the curvature of the potential energy of each atom, evaluated at its equilibrium position, is assumed to be positive definite, and therefore the system is mechanically stable. Thus no diffusion is allowed and the theory is therefore inapplicable to liquids, or to those solids or molecules where there are two adjacent minima in the potential energy (as a function of some coordinate in the system) separated by an energy barrier comparable to kT, no matter how close those minima are in configurational space. Such double potential wells are important in the low temperature thermodynamics of amorphous solids, and they may also exist at crystalline defects such as grain boundaries.

Rewriting the free energy

In general, it is necessary to make an approximation in order to write down the free energy of the system as a function of the average atomic positions. In the harmonic approximation the potential energy is expanded to second order in the displacements of the atoms from their mean positions. Let $u_{i\alpha}$ be the displacement of atom i in the α direction ($\alpha = x, y$ or z) from the equilibrium positions $\boldsymbol{r}_i$. The potential energy is given by

$$E_{\mathrm{p}} = E_{\mathrm{p}}(\boldsymbol{r}_1, \ldots, \boldsymbol{r}_N) - \sum_{i\alpha} f_{i\alpha}\, \delta u_{i\alpha} + \tfrac{1}{2} \sum_{i\alpha} \sum_{j\beta} D_{i\alpha j\beta}\, u_{i\alpha}\, u_{j\beta}, \tag{2.1}$$

where $f_{i\alpha} = -\partial E_{\mathrm{p}}/\partial r_{i\alpha}$ and $D_{i\alpha j\beta} = \partial^2 E_{\mathrm{p}}/\partial r_{i\alpha}\partial r_{i\beta}$ are evaluated at the equilibrium positions. It follows that $f_{i\alpha}$ are all zero. The potential energy E_{p} is expressed either in terms of some interatomic potential, such as a sum of pair potentials or N-body potentials, or as the potential energy surface defined by an electronic structure calculation. Writing $u_{i\alpha}(t)$ as $u_{i\alpha}\,e^{i\omega t}$, the equations of motion may be expressed as follows:

$$\omega^2 \tilde{u}_{i\alpha} = \sum_{j\beta} \tilde{D}_{i\alpha j\beta}\, \tilde{u}_{j\beta}, \tag{2.2}$$

where $\tilde{u}_{i\alpha}$ is equal to $\surd(m_i)\, u_{id}$ and $\tilde{D}_{i\alpha j\beta}$ is equal to $D_{i\alpha j\beta}/\surd(m_i\, m_j)$, and m_i is the mass of atom i. The eigenvalues ω_n^2 may be determined by solving the secular determinant $\det(\omega^2 I - \tilde{D}) = 0$ and the normalized eigenvectors $\tilde{u}_{i\alpha}^{(n)}$ may then be obtained from (2.2) by setting ω^2 equal to ω_n^2. The total density of states is given by $Y(\omega^2) = \Sigma_n\, \delta(\omega^2 - \omega_n^2)$. The Helmholtz free energy is given by

$$F = E_{\mathrm{p}} + kT \int_0^\infty N(\omega) \ln\left[2 \sinh\left(\frac{h\omega}{4\pi kT}\right)\right] \mathrm{d}\omega, \tag{2.3}$$

where the density of states $N(\omega)$ is related to the density of states $Y(\omega^2)$ by $Y(w^2) = N(\omega)/2\omega$. The equilibrium position i is given by the condition that $-\nabla_i F = 0$. From (2.3) for the free energy we see that there are two contributions to $-\nabla_i F$. The first is the temperature independent force due to the potential energy $-\nabla_i E_{\mathrm{p}}$. This is the force that we would consider in the absence of any vibratory motion of the atoms. The second contribution, $-\nabla_i(F - E_{\mathrm{p}})$, is temperature dependent and arises from the fact that the density of states changes as atom i undergoes a virtual displacement because elements of the matrix $\tilde{D}$ change. This contribution is not zero even when $T = 0$ owing to the zero point motion of the atoms. The changes in the matrix $\tilde{D}$ as an atom is displaced are due to the anharmonicity of the potential energy E_{p}. Thus, although the expansion of the potential energy is carried out to only second order in the atomic displacements, the matrix elements $D_{i\alpha j\beta}$ vary as the equilibrium atomic positions change because of higher order derivatives of the potential energy. For this reason we are really discussing quasiharmonic theory.

Expressions of the temperature dependent force

Approximate solutions for $-\nabla_i(F - E_{\mathrm{p}})$ have appeared (Sutton 1989; LeSar *et al.* 1989), but before we outline them we give the *exact* solution. Using (2.2) and (2.3) it may be shown that

$$-\nabla_i(F - E_{\mathrm{p}}) = -\operatorname{tr} \rho \nabla_i \tilde{D}, \tag{2.4}$$

where tr denotes trace and the elements of the matrix ρ are

$$\rho_{j\beta k\alpha} = \tfrac{1}{2} \sum_n \frac{E(\omega_n)}{\omega_n^2}\, \tilde{u}_{j\beta}^{(n)}\, \tilde{u}_{k\alpha}^{(n)*}. \tag{2.5}$$

The sum over n is taken over all normal modes of the system and $E(\omega_n)$ is the internal energy associated with the nth normal mode:

$$E(\omega_n) = \frac{h\omega_n}{2\pi}\left(\frac{1}{2} + \frac{1}{\exp(h\omega_n/2\pi kT) - 1}\right). \tag{2.6}$$

It is clear from (2.4) that $-\nabla_i(F - E_{\mathrm{p}})$ is non-zero only if the potential has non-zero third derivatives and is therefore anharmonic. The force diverges if any normal mode frequency approaches zero, which indicates that the system must be mechanically stable to avoid such singularities. In (2.4) the $3N \times 3N$ matrix $\tilde{D}$ must be diagonolized where N is the number of atoms in the cluster. If periodic boundary conditions are used N is the number of atoms in the unit cell and the matrix D has to be diagonolized at an appropriate number of $\boldsymbol{k}$-points in the Brillouin zone. The more approximate methods outlined below were developed to avoid the diagonalization of such large matrices. In the classical limit where $kT \gg h\omega/2\pi$ the free energy equation (2.3)) becomes

$$F_{\text{class}} = E_{\text{p}} + 3kT \ln\left[\left(\frac{h}{2\pi kT}\right)^{N} |\tilde{D}|^{\frac{1}{6}}\right] \tag{2.7}$$

and eqn. (2.4) becomes

$$-\nabla_i(F_{\text{class}} - E_{\text{p}}) = -\sum_n \tfrac{1}{2}kT\nabla_i \ln \omega_n^2 = -\tfrac{1}{2}kT\nabla_i \ln|\tilde{D}|, \tag{2.8}$$

where $|\tilde{D}|$ is the determinant of the $3N \times 3N$ matrix $\tilde{D}$.

Einstein models

LeSar *et al.* (1989) approximate (2.7) with an Einstein model in which $|\tilde{D}|$ becomes a product of N 3×3 determinants, $|\tilde{D}_k|$, one for each atom. They called this the 'local harmonic model', but I feel 'classical Einstein model' is more apposite. The matrix elements $\tilde{D}_{k\alpha k\beta}$ are found from the condition that the energy of the system is invariant with respect to a rigid translation:

$$D_{k\alpha k\beta} = -\sum_{j\neq k} D_{k\alpha j\beta}. \tag{2.9}$$

Thus the Einstein frequencies are determined by interactions with neighbouring sites (note the absence of tildas above the matrix elements in (2.9)), even though there is no dispersion in the Einstein approximation. In this approximation the free energy of the system in the classical limit becomes

$$F^{\text{E}}_{\text{class}} = E_{\text{p}} + 3kT \sum_{j=1}^{N} \ln\left[\frac{h|\tilde{D}_j|^{\frac{1}{6}}}{2\pi kT}\right]. \tag{2.10}$$

The equilibrium position of atom i is determined by $-\nabla_i F^{\text{E}}_{\text{class}} = 0$, which is now straightforward to evaluate:

$$-\nabla_i F^{\text{E}}_{\text{class}} = -\nabla_i E_{\text{p}} - \tfrac{1}{2}kT\sum_j \nabla_i \ln|\tilde{D}_j|. \tag{2.11}$$

LeSar *et al.* (1989) compared the estimation of the Helmholtz free energy of a perfect Cu crystal using the above classical Einstein model with a Monte Carlo procedure in which the quasiharmonic approximation was not made. The copper crystal was modelled by a pairwise Morse potential truncated between the second and third neighbours. The two sets of results are indistinguishable. They also compared the vacancy formation free energy in the two methods. The errors range from 0 to 1.2% as the temperature is increased from zero up to about 75% of the melting point of the model for Cu. This agreement is perhaps surprisingly good in view of the approximations that are made in the classical Einstein model. Both the classical Einstein model and the Monte Carlo procedure yield incorrect free energies at temperatures below the Debye temperature of the model because they neglect the quantum freezing out of modes.

Najafabadi *et al.* (1991 *a*) applied the classical Einstein model with embedded atom potentials for gold (Foiles *et al.* 1986) to simulate the structural evolution and excess thermodynamic properties of twelve (001) twist boundaries for temperatures between 0 and 700 K. Four of the twelve grain boundaries underwent first-order structural phase transitions as seen by the crossing of the free energy against temperature curves for the competing structures. The grain boundary linear thermal expansion coefficient varied with misorientation in a similar way to the excess grain boundary entropy.

Approaches based on the local atomic environment

Sutton (1989) developed a different strategy in which neither the classical limit nor the Einstein approximation are assumed. Returning to the Helmholtz free energy in (2.3) we can always write the total density of states $N(\omega)$ exactly as a sum of local densities of states:

$$N(\omega) = \sum_{j=1}^{N} n_j(\omega) = 2\omega \sum_{j=1}^{N} y_j(\omega^2), \tag{2.12}$$

where the local density of states $y_j(\omega^2)$ is defined by

$$y_j(\omega^2) = \sum_n |\tilde{u}_j^{(n)}|^2 \delta(\omega^2 - \omega_n^2), \tag{2.13}$$

and the sum is over all $3N$ normal modes. The Einstein model approximates $y_j(\omega^2)$ by three delta functions. The strategy taken by Sutton (1989) was to approximate the local density of states by using information about the local atomic environment through the moments theorem and the known functional form of the local density of states at the band edges. By fitting more and more moments of the local density of states we obtain increasingly accurate approximations to the true density of states.

Let $M_j^{(p)}$ denote the pth moment of the local density of states $y_j(\omega^2)$:

$$M_j^{(p)} = \int_0^\infty y_j(\omega^2)\, \omega^{2p}\, \mathrm{d}\omega^2 = \int_0^\infty n_j(\omega)\, \omega^{2p}\, \mathrm{d}\omega = \mu_j^{(2p)}. \tag{2.14}$$

Thus, $M_j^{(p)}$ is equal to the $2p$th moment, $\mu_j^{(2p)}$, of the local density of states $n_j(\omega)$. Using the moments theorem (Cyrot-Lackmann 1968) the second moment of $n_j(\omega)$ is given by

$$\mu_j^{(2)} = M_j^{(1)} = \sum_{\alpha=1}^{3} \tilde{D}_{j\alpha j\alpha} = \nabla_j^2 E_\mathrm{p}/m_j. \tag{2.15}$$

The first moment $M_j^{(1)}$ was fitted to an assumed function form for the local density of states $y_j(\omega^2)$. In a three dimensional crystal the density of states must vary like the square root of ω^2 at the band edges. The lower band edge is always at $\omega^2 = 0$, and the integral of $y_j(\omega^2)$ over the whole band must equal 3. The simplest choice of functional form for $y_j(\omega^2)$, satisfying these constraints, is the following:

$$y_j(\omega^2) = (6/\pi M_j^{(1)})\,[(M_j^{(1)})^2 - (\omega^2 - M_j^{(1)})^2]^{\frac{1}{2}}, \tag{2.16}$$

This form is a semi-elliptic density of states, which is non-zero between $\omega^2 = 0$ and $2M_j^{(1)}$, with the centre of gravity at $M_j^{(1)}$. The corresponding local density of states $n_j(\omega)$ is given by

$$n_j(\omega) = \frac{12\omega^2}{\pi(\mu_j^{(2)})^2}(2\mu_j^{(2)} - \omega^2)^{\frac{1}{2}}, \tag{2.17}$$

which is proportional to ω^2 at low frequencies. At $T = 0$ K the Helmholtz free energy differs from the potential energy due to zero point motion. Taking the limit of $T=0$ in (2.3) we obtain

$$F_{\text{zero}} = E_{\text{p}} + \frac{h}{4\pi}\sum_j \mu_j^{(1)} \approx E_{\text{p}} + 1.44\frac{h}{2\pi}\sum_j [\mu_j^{(2)}]^{\frac{1}{2}}. \tag{2.18}$$

Thus, at $T=0$ K the Helmholtz free energy has the form of a Finnis–Sinclair potential (Finnis & Sinclair 1984), with the square root term arising from zero point motion. In an Einstein model the zero point energy of atom j is $1.5h\omega_j^{\text{E}}/2\pi$. From (2.18) it follows that $\omega_j^{\text{E}} = 0.96[\mu_j^{(2)}]^{\frac{1}{2}}$.

At intermediate temperatures, where quantum freezing out of modes is still important, we must use the full form of the Helmholtz free energy, equation (2.3). In our second moment approximation this becomes

$$F = E_{\text{p}} + \sum_{i=1}^{N} F_i, \tag{2.19}$$

where
$$F_i = \frac{48kT}{\pi}\int_0^1 x^2(1-x^2)^{\frac{1}{2}} \ln[2\sinh(\tfrac{1}{2}c_i x)]\,\mathrm{d}x, \tag{2.20}$$

and
$$c_i = h[2\mu_i^{(2)}]^{\frac{1}{2}}/2\pi kT. \tag{2.21}$$

The free energy in (2.19) comprises the potential energy E_{p} and a sum of *projections* F_i of the vibrational free energy of the whole system onto individual sites. The projection is effected by the local densities of states, which are projections of the global density of states onto individual atomic sites. When c_i is infinite the temperature is zero and F reduces to F_{zero} given in (2.18). At the other extreme limit where c_i tends to zero we obtain the classical limit, which in our second moment model is given by

$$F_i^{\text{class}} = 3kT[\ln(\tfrac{1}{2}c_i) + \tfrac{1}{4}]. \tag{2.22}$$

Quantum effects begin to be important when $c_i \geqslant 1$. The temperature $\theta_i = h[2\mu_i^{(2)}]^{\frac{1}{2}}/(2\pi k)$ has the meaning of a local Debye temperature.

The vibrational entropy, internal energy and specific heat may be projected onto individual atomic sites (Sutton 1989). Expressions were also given for the mean square displacement $\langle u_i^2 \rangle$ and a local Grüneisen parameter γ_i. The effective force acting on atom i is given by

$$-\nabla_i F = -\nabla_i E_{\text{p}} - \sum_j \tfrac{1}{2} U_j \nabla_i (\ln \nabla_j^2 E_{\text{p}}), \tag{2.23}$$

where U_j is the internal energy projected onto site j, which is given by

$$U_j = \frac{24kT}{\pi} c_j \int_0^1 x^3(1-x^2)^{\frac{1}{2}} \coth(\tfrac{1}{2}c_j x)\,\mathrm{d}x. \tag{2.24}$$

Sutton (1989) applied this second moment model to a study of the thermodynamic properties of the 22.06° twist boundary in gold. By far the most significant structural change with increasing temperature was the increase in the boundary expansion. A strong correlation (over 90%) was found between the local stiffness parameter c_i, equation (2.21) and the local hydrostatic pressure. Thus, compressed sites are associated with low contributions to the excess vibrational entropy, specific heat and mean square displacement and high (positive) contributions to the excess vibrational free energy and internal energy. Compressed sites are also well correlated (over 90%) with low Grüneisen constants.

Comparison of approaches

Comparing (2.23) with the force in the classical Einstein model, equation (2.11), we see that the two are very similar in that the temperature dependent contribution is an N-body force which depends on the third derivatives of the potential energy. However, the determinant of $\tilde{D}_j$ is used in the local Einstein model, whereas the trace of $\tilde{D}_j$ is used in the second moment model. Thus, the off-diagonal elements of the 3×3 matrix $\tilde{D}_j$ are taken into account in the local Einstein model, but not in the second moment model. The off-diagonal elements of $\tilde{D}_j$ describe the resistance of the local atomic environment to shear. Local shears are important modes of thermal excitation in open crystal structures (Barron *et al.* 1980). Local shears parallel to the plane of a grain boundary are also known to be important local modes of thermal excitation. A model that combines all the information contained in each matrix $\tilde{D}_j$ with a continuous density of states $y_j(\omega^2)$, displaying square root singularities at the band edges, would be preferable to both the classical Einstein model and the second moment model. This may be achieved by replacing the three delta functions representing the local density of states in the classical Einstein model by three semi-elliptic bands, one for each Einstein mode. The effective force on atom j becomes

$$-\nabla_j F = -\nabla_j E_{\mathrm{p}} - \sum_{i=1}^{N}\sum_{\nu=1}^{3}\frac{U_{i\nu}}{2\omega_{i\nu}^2}\sum_{\alpha=1}^{3}\sum_{\beta=1}^{3} x_{i\beta}^{(\nu)}\, x_{i\alpha}^{(\nu)*}\,\nabla_j \tilde{D}_{i\alpha i\beta}, \tag{2.25}$$

where $\omega_{i\nu}^2$ are the eigenvalues and $x_{i\alpha}^{(\nu)}$ are the eigenvectors of the 3×3 matrix $\tilde{D}_i$ and $U_{i\nu}$ is the internal energy associated with the νth Einstein mode at atom i. In the high temperature limit, where $U_{i\nu}$ becomes kT, we recover (2.11) of the classical Einstein model. But at lower temperatures, where quantum effects reduce $U_{i\nu}$, we obtain a more accurate description of the forces and thermodynamic functions than in either the local Einstein model or the second moment model of Sutton (1989). Comparing (2.25) with the exact expression for the quasiharmonic effective force, (2.4), we see that the off-diagonal elements ($i \neq j$) of the matrix ρ are ignored in (2.25). The incorporation of such intersite correlations would entail taking into account higher moments of the local densities of states.

3. Direct free energy minimization in a substitutional alloy

Consider a bicrystal containing A and B atoms. The interface may be of the homophase type, e.g. a grain boundary separating misoriented crystals of an AB alloy or an anti-phase boundary separating two ordered AB alloy crystals, or of the heterophase type, e.g. an interface separating crystals of (initially) pure A and pure B. Whatever the interface type it is assumed that a fixed number of atomic sites exists, which limits the treatment to substitutional alloys. It is not assumed that the positions of the atomic sites are fixed. The problem at hand is to determine the equilibrium distribution of A and B atoms in the vicinity of the interface, at a given temperature and pressure. The problem is still not fully defined until we specify whether the numbers of A and B atoms are fixed or whether the chemical potentials of A and B atoms are fixed. We assume the latter, since the crystals on either side of the interface act as large reservoirs of A and B atoms, at fixed chemical potentials μ_{A} and μ_{B}, which can exchange atoms with the interfacial region.

Even at equilibrium there are fluctuations in the atomic structure and solute distribution. Thus, if we took two snapshots of the interface at different instants we might see different positions of solute atoms. But if we averaged the position and occupancy of each site over a large period of time we would expect, at equilibrium, to obtain convergent values. Thus the problem at hand addresses the time averaged atomic structure and solute distribution within the interface. The time averaged quantities, at equilibrium, are the expectation values for those quantities computed in the appropriate statistical ensemble. The present ensemble is a reduced grand canonical ensemble since the numbers of A and B atoms are not fixed, but no vacant sites are allowed. In principle we may regard the vacancies as a third alloy component, and thereby recover the grand canonical ensemble, but in practice there are difficulties associated with vacancies as described below.

Expressions for thermodynamic quantities

Let p_i denote the occupancy of site i at a given instant in time. We set p_i equal to 1(0) if site i is occupied by a B(A) atom. Let the ensemble average of p_i be $\langle p_i \rangle = c_i$. We call c_i the occupancy of site i, with the understanding that it means the average occupancy of the site at equilibrium. The site occupancy is a number lying between 0 and 1. Let $\boldsymbol{R}_i$ be the position of site i at any given instant and let the time average of $\boldsymbol{R}_i$ be $\langle \boldsymbol{R}_i \rangle = \boldsymbol{r}_i$. Our task is to find the site occupancies $\{c_i\}$ and positions $\{\boldsymbol{r}_i\}$ which minimize the grand potential of the system. We shall write down an expression for the grand potential in terms of the sets $\{c_i\}$ and $\{\boldsymbol{r}_i\}$ and demand that it is minimized with respect to these variables. These ideas first appeared in Gyorffy & Stocks (1983) and Lundberg (1987).

We express the grand potential, Ω, as follows:

$$\Omega = F(\{c_i\}, \{\boldsymbol{r}_i\}) - TS_c - \mu_A N_A - \mu_B N_B. \tag{3.1}$$

F is the Helmholtz free energy of the ensemble excluding the configurational entropy:

$$F(\{c_i\}, \{\boldsymbol{r}_i\}) = U(\{c_i\}, \{\boldsymbol{r}_i\}) - TS_v(\{c_i\}, \{\boldsymbol{r}_i\}), \tag{3.2}$$

where U is the internal energy, including the vibrational contribution, and S_v is the vibrational entropy. In (3.1) the configurational entropy is denoted by S_c. In the Bragg–Williams approximation for S_c we assume that the configurational entropy is that of an ideal (i.e. non-interacting) mixture of A and B atoms. This is an upper bound because the interactions introduce correlations between the occupancies of the sites which reduce the configurational entropy. In this approximation S_c is given by

$$S_c = -k \sum_i c_i \ln c_i + (1-c_i) \ln (1-c_i), \tag{3.3}$$

N_A and N_B are the numbers of A and B atoms:

$$N_A = \sum_i (1-c_i), \quad N_B = \sum_i c_i. \tag{3.4}$$

Inserting (3.2)–(3.4) into (3.1) for Ω and minimizing with respect to c_k we obtain

$$\frac{\partial F}{\partial c_k} + kT \ln \left(\frac{c_k}{1-c_k} \right) = \mu_B - \mu_A. \tag{3.5}$$

At equilibrium, therefore, the local chemical potential difference, $\mu_B^k - \mu_A^k$, which equals the left-hand side of (3.5), is the same at all sites and equal to $\mu_B - \mu_A$. The equilibrium condition involves the difference in chemical potentials $\mu_B - \mu_A$ because, in a substitutional alloy, equilibrium is attained by exchanging atoms between sites. The variation of the occupancy from site to site at a defect is a result of the variation of $\partial F/\partial c_k$. If $\partial F/\partial c_k \gg (\mu_B - \mu_A)$ then site k is occupied by an A atom almost all the time, and, conversely if $\partial F/\partial c_k \ll (\mu_B - \mu_A)$ site k is occupied by a B atom almost all the time.

Role of configurational entropy

To illustrate the role of the configurational entropy let us ignore the vibrational entropy contribution so that the Helmholtz free energy, $F(\{c_i\}, \{\boldsymbol{r}_i\})$, reduces to the potential energy. We assume the potential energy may be represented by a sum of pair potentials. We denote the energy of interaction between an A atom at site i and a B atom at site j by $\epsilon_{ij}^{AB} = \epsilon_{ji}^{AB}$, with the understanding that ϵ_{ij}^{AB} is a function of $|\boldsymbol{R}_i - \boldsymbol{R}_j|$. Similar symbols are used for A–A and B–B interaction. The hamiltonian of the system is given by

$$\mathscr{H} = \sum_{\substack{i,j \\ i \neq j}} p_i p_j \theta_{ij} + \sum_i (p_i \alpha_i + d_i^A), \tag{3.6}$$

where

$$\left.\begin{aligned} \theta_{ij} &= \tfrac{1}{2}[\epsilon_{ij}^{BB} + \epsilon_{ij}^{AA} - 2\epsilon_{ij}^{AB}], \\ \alpha_i &= \sum_{j \neq i} (\epsilon_{ij}^{AB} - \epsilon_{ij}^{AA}) - (\mu_B - \mu_A), \\ d_i^A &= (\tfrac{1}{2} \sum_{j \neq i} \epsilon_{ij}^{AA}) - \mu_A. \end{aligned}\right\} \tag{3.7}$$

The connection between the grand potential and the hamiltonian is $\Omega = -kT \ln Z$ where Z is the grand partition function, $Z = \mathrm{tr}\, \mathrm{e}^{-\beta \mathscr{H}}$, and $\beta = 1/kT$. To evaluate the trace we imagine the atom positions are temporarily frozen and consider all the states of the system characterized by the set of integers $p_1, p_2, p_3, \ldots$, each of which can be zero or one. We can obtain a useful identity for the expectation value $\langle p_k \rangle = c_k = \mathrm{tr}\, p_k \mathrm{e}^{-\beta \mathscr{H}}/Z$ by evaluating the trace in a particular way due to Callen (1963). The result (Balcerzak 1991) is

$$c_k = \langle p_k \rangle = \langle (1 + \mathrm{e}^{\beta \gamma_k})^{-1} \rangle, \tag{3.8}$$

where γ_k is the 'local field' at site k:

$$\gamma_k = \alpha_k + 2 \sum_{j \neq k} p_j \theta_{jk}. \tag{3.9}$$

Equation (3.8) is useful because it is an identity against which standard approximations such as mean field theory and the auto-correlation approximation (AA) may be tested. In the mean field approximation (MFA) correlations between the occupancies on different sites are ignored and it is also assumed that $\langle p_j^n \rangle = \langle p_j \rangle^n$. In that case c_k becomes

$$c_k = (1 + \mathrm{e}^{\beta \gamma_k^{\mathrm{MFA}}})^{-1} \tag{3.10}$$

where

$$\gamma_k^{\mathrm{MFA}} = \alpha_k + 2 \sum_{j \neq k} c_j \theta_{jk}. \tag{3.11}$$

The physical meaning of the local field is that it is the difference between the energy of replacing an A atom at site k with a B atom, and $(\mu_B - \mu_A)$. Comparing with the exact result (equations (3.8) and (3.9)), it is seen that the $\langle .. \rangle$ brackets have been taken inside the exponent in the MFA. The grand partition function in the MFA, Z^{MFA}, is readily evaluated for a given set of atomic positions and the following transparent form for the grand potential, Ω^{MFA}, may thus be obtained:

$$\Omega^{\mathrm{MFA}} = \tfrac{1}{2} \sum_{\substack{i,j \\ i \neq j}} (c_i c_j \epsilon_{ij}^{\mathrm{BB}} + [c_i(1-c_j) + c_j(1-c_i)]\, \epsilon_{ij}^{\mathrm{AB}} + (1-c_i)(1-c_j)\, \epsilon_{ij}^{\mathrm{AA}})$$
$$+ kT \sum_i [c_i \ln c_i + (1-c_i) \ln (1-c_i)] - \mu_{\mathrm{A}} \sum_i (1-c_i) - \mu_{\mathrm{B}} \sum_i c_i. \quad (3.12)$$

The first term is the internal energy, which is obtained by replacing the site occupancy operators in the Hamiltonian, equation (3.6), by their expectation values. The second term is the contribution from the configurational entropy. The final term is the contribution from the chemical potentials of the A and B atoms. Thus, in the MFA each A or B atom is replaced by a *hybrid atom*, which varies in the degree of its A property or B property with the local atomic environment in a continuous and self-consistent manner. It is easily shown that minimization of the grand potential with respect to c_k leads to (3.10). We may think of $-\partial \Omega / \partial c_i$ as a generalized force, conjugate to the site occupancy c_i. But in (3.10) the grand potential is also a function of all the atomic coordinates in the system, through the pair potentials. We should, therefore, minimize the grand potential with respect to all $4N$ variables, where N is the total number of sites. In this way the atomic structure of the interface changes as the degree of segregation changes, for example because the chemical potential difference $\mu_{\mathrm{B}} - \mu_{\mathrm{A}}$ is changed.

Surface and interface segregation

Najafabadi *et al.* (1991*b*) used embedded atom potentials (Foiles *et al.* 1986) in a MFA to model surface and interfacial segregation in a Cu–Ni solid solution alloy, and they included the vibrational entropy contribution in the classical Einstein model. The atomic mass associated with site j was set equal to $c_j m_{\mathrm{B}} + (1-c_j) m_{\mathrm{A}}$, which is consistent with the notion that each site is occupied by a hybrid atom. The results of the simulations for the surface segregation profiles compared very well, layer by layer, with Monte Carlo simulations, using the same interatomic potentials, where the mean field, Bragg–Williams and classical Einstein approximation are not made. Three (001) twist boundaries were also studied with bulk alloy copper concentrations of 10, 50 and 90 at. %. The only significant disagreement with the Monte Carlo simulations was at the low bulk copper concentration, where the concentration of copper in the boundary plane was overestimated by 59% in the 10.4° boundary. However, the *trends* in the segregation profiles of the Monte Carlo results are well reproduced in all boundaries studied. Not only is this new approach much faster than Monte Carlo methods it enables excess thermodynamic state functions to be determined as an important by-product, and it offers much greater physical insight into the local environmental factors driving segregation.

If the composition of the interface is predicted to be quite different from that of the bulk, or if an ordered phase is predicted at the interface while the adjoining crystals remain compositionally disordered, the configurational entropy of the system will be reduced. In that case, since the Bragg–Williams approximation overestimates the configurational entropy, it may be said that the configurational entropy of the system is lower despite the Bragg–Williams approximation. The

Bragg–Williams approximation is therefore not a reason for concern except in those cases where the configurational entropy is predicted to increase, such as the compositional disordering of a grain boundary, or fault, in an ordered alloy.

Role of correlations

The MFA is almost certainly more severe. Not only does this approximation ignore correlations in the occupancies of different sites but it also ignores self-correlations. Experience in magnetic systems (Balcerzak *et al.* 1990), indicates that inclusion of self-correlations gives the largest correction to the MFA. Self-correlations refer to the fact that $\langle p_j^n \rangle = \langle p_j \rangle$, rather than $\langle p_j^n \rangle = \langle p_j \rangle^n$ which is assumed in mean field theory. Using the pair potential model described above it is possible to include these self-correlations while continuing to ignore correlations on different sites. This is known as the auto-correlation approximation (AA) (see Balcerzak (1991) for a thorough discussion of the magnetic case).

Let us reconsider the grand partition function, and $e^{-\beta \mathscr{H}}$ in the pair potential model, (3.6). It is useful to rewrite $\exp(-\beta\theta_{ij} p_i p_j)$ as follows:

$$\exp(-\beta\theta_{ij} p_i p_j) = 1 + p_i[\exp(-\beta\theta_{ij} p_j) - 1], \tag{3.13}$$

where we have used $p_i^n = p_i$. Inserting this expression for $\exp(-\beta\theta_{ij} p_i p_j)$ into $e^{-\beta \mathscr{H}}$ we obtain

$$e^{-\beta \mathscr{H}} = \prod_n \exp(-\beta(p_n \alpha_n + c_n^{\mathrm{A}})) \prod_i \prod_{j \neq i} (1 - p_i + p_i \exp(-\beta\theta_{ij} p_j))]. \tag{3.14}$$

This equation is still exact. In the MFA term $\exp(-\beta\theta_{ij} p_j)$ is replaced by $\exp(-\beta\theta_{ij} \langle p_j \rangle) = \exp(-\beta\theta_{ij} c_j)$. In the AA it is replaced by

$$\left.\begin{aligned} \langle \exp(-\beta\theta_{ij} p_j) \rangle &= \langle 1 + p_j(\exp(-\beta\theta_{ij}) - 1) \rangle \\ &= 1 + c_j(\exp(-\beta\theta_{ij}) - 1) = \exp -(\beta\xi_{ij} c_j), \end{aligned}\right\} \tag{3.15}$$

where ξ_{ij} is defined by

$$\xi_{ij} = -(kT/c_j)\ln[1 + c_j(\exp(-\beta\theta_{ij}) - 1)]. \tag{3.16}$$

We can think of the ξ_{ij} as effective interaction parameters which result from the improved treatment of the statistical averaging. Note that as $c_j \to 0$ then $\xi_{ij} \to kT(1 - e^{-\beta\theta_{ij}})$, which tends to θ_{ij} as $T \to 0$. Also, as $c_j \to 1$ then $\xi_{ij} \to \theta_{ij}$. But for $0 < c_j < 1$ the effective pair potential ξ_{ij} is less than θ_{ij}. Clearly, the effective interaction parameters are dependent on temperature and the local concentration.

The 'local field', equation (3.11), the self-consistency condition (3.10), and the grand potential equation (3.12), are precisely the same in the auto-correlation approximation as in the MFA, except that where the interaction parameters θ_{ij} appear in the MFA they are replaced by the effective interaction parameters ξ_{ij}. We conclude that the only difference between the MFA and AA is the change in definition of the local field, with the interaction parameters θ_{ij} being replaced by effective interaction parameters ξ_{ij}, which are temperature and concentration dependent.

Role of relaxation

Both the MFA and AA are expected to break down when there is extensive relaxation around solute atoms, due to a large size misfit for example, especially in the limit of small concentrations. (In practice the same comment applies also to Monte Carlo simulations but for different reasons.) For example, consider a single face centred cubic (FCC) crystal of A atoms containing vacant sites, in which atomic interactions are modelled by pair potentials ϵ_{ij}^{AA}. In this case the 'alloy' is between A atoms and vacancies and it is disordered. Rather than treating each site as being occupied ($p_i = 0$) or vacant ($p_i = 1$) the MFA treats all sites in the crystal as having the same vacancy occupancy c, which is determined by the following self-consistency condition:

$$c = 1/[1+\exp(\beta(\mu_A-(1-c)\sum_{j\neq 1}\epsilon_{ij}^{AA}))]. \quad (3.17)$$

(Although we are discussing the MFA here our remarks apply equally to the AA.) We have assumed that the vacancies are in thermal equilibrium so that their chemical potential is zero. The failure of the MFA in this case lies in the inadequacy of its description of the relaxation around each vacancy. In reality there is relaxation around each vacant site in the crystal which affects the formation energy to a significant degree. This relaxation renders sites that are occupied by atoms non-equivalent. The only form of relaxation that appears in the mean field theory is that the lattice parameter of the FCC crystal is altered very slightly, and all sites remain equivalent. Thus the mean field treatment essentially neglects the relaxation energy of each vacancy. Similarly the relaxation energy in any dilute AB alloy is underestimated in the MFA. In such cases the approximation will be most successful where the relaxation energy is small.

Part of §3 grew out of discussions with T. Balcerzak. This research was supported by the EC under contrast No. SC1*–CT91–0703 (TSTS).

References

Balcerzak, T. 1991 Cluster identities in localized spin systems. *J. Magn. magn. Mater.* **97**, 152–168.

Balcerzak, T., Mielnicki, J., Wiatrowski, G. & Urbaniak-Kucharczyk, A. 1990 The magnetisation and the correlation functions in thin diluted film (Ising model, $S = \frac{1}{2}$). *J. Phys. Condensed Matter* **2**, 3955–3966.

Barron, T. H. K., Collins, J. G. & White, G. K. 1980 Thermal expansion of solids at low temperatures. *Adv. Phys.* **29**, 609–730.

Callen, H. B. 1963 A note on Green functions and the Ising model. *Phys. Lett.* **4**, 161.

Cyrot-Lackmann, F. 1968 Sur le calcul de la cohésion et de la tension superficielle des métaux de transition par une méthode de liaison fortes. *J. Phys. Chem. Solids* **29**, 1235–1243.

Finnis, M. W. & Sinclair, J. E. 1984 A simple empirical N-body potential for transition metals. *Phil. Mag.* A **50**, 45–55.

Foiles, S. M., Daw, M. S. & Baskes, M. I. 1986 Embedded-atom-method functions for the fcc metals Cu, Ag, Au, Ni, Pd, Pt and their alloys. *Phys. Rev.* B **33**, 7983–7991.

Gyorffy, B. L. & Stocks, G. M. 1983 Concentration waves and Fermi surfaces in random metallic alloys. *Phys. Rev. Lett.* **50**, 374–377.

LeSar, R., Najafabadi, R. & Srolovitz, D. J. 1989 Finite-temperature defect properties from free-energy minimization. *Phys. Rev. Lett.* **63**, 624–627.

Lundberg, M. 1987 Surface segregation and relaxation calculated by the embedded-atom method: application to face-related segregation on platinum-nickel alloys. *Phys. Rev.* B **36**, 4692–4699.

Najafabadi, R., Srolovitz, D. J. & LeSar, R. 1991*a* Thermodynamic and structural properties of [001] twist boundaries in gold. *J. mater. Res.* **6**, 999–1011.

Najafabadi, R., Wang, H. Y., Srolovitz, D. J. & LeSar, R. 1991*b* A new method for the simulation of alloys: application to interfacial segregation. *Acta metall. Mater.* **39**, 3071–3082.

Rickman, J. M. & Phillpot, S. R. 1991 Calculation of the 'absolute' free energy and the entropy of classical solids from the motion of particles. *J. chem. Phys.* **95**, 7562–7568.

Sutton, A. P. 1989 Temperature-dependent interatomic forces. *Phil. Mag.* A **60**, 147–159.
Sutton, A. P. 1991 An analytic model for grain boundary expansions and cleavage energies. *Phil. Mag.* A **63**, 793–818.

Discussion

P. A. Mulheran (*Harwell Laboratory, Didcot, U.K.*): As I understand it the Srolovitz method (and presumably your method also) is to calculate a local frequency to represent a whole phonon branch. When the atomic configuration is altered the phonon branches change and these changes are mirrored in the local frequencies. The way these frequencies are calculated by the Srolovitz group is reasonable provided that the interatomic forces are short-ranged and I assume that this comment also applies to your calculation. However if the forces are long-ranged as in ionic systems the correlations between the moving ions must be considered since long-range polarization fields contribute to the vibrational frequencies. This necessarily precludes the use of entirely localized modes. Nevertheless we can use the same concept provided that the ionic system has two-dimensional symmetry; we cannot work with point or line defects.

A. P. Sutton: The idea of both the local Einstein and the second moment models is to use information about local atomic environment to characterize the local vibrational spectrum. The information that is incorporated consists of certain local *second* derivatives of the potential energy, which, as you say, are often longer-ranged than the first derivatives of the potential energy.

Perhaps a greater concern might be the *averaging* over all the normal modes of the system that is done in both of these simple models. If particular modes dominate the vibrational properties of a defect then one would like to see them treated explicitly. This is done only in the exact expression for the force (within the quasiharmonic approximation) given in (2.4).

A. M. Stoneham (*Harwell Laboratory, Didcot, U.K.*): I think your exact theory could be used to check a further model. Temperature-dependent empirical potentials (i.e. refitted to crystal properties at each temperature) are used for defect studies. It is not clear that these are fully transferrable, and a check would be useful.

A. P. Sutton: Yes the exact formula, given in (2.4), would enable the consistency of the potentials fitted to properties measured at different temperatures to be tested.

Theory of impurity and defect induced instabilities

By D. J. Chadi

NEC Research Institute, 4 Independence Way, Princeton, New Jersey 08540-6620, U.S.A.

A brief review of the current status of our work on substitutional and interstitial impurities and native defects in GaAs and of the modelling process used in the calculations is given. The combined empirical tight-binding and *ab initio* pseudopotential approach utilized in these studies allows for efficient testing of a large number of structural possibilities and the identification of the most relevant ones. Applications of the method and new results for EL2, DX, and self-interstitial defect centres in GaAs are discussed.

1. Introduction

It is now well known that simple point defects and impurities in a III–V semiconductor such as GaAs can give rise to structural instabilities that profoundly affect their electronic and optical properties. In many cases the instability is driven by charge exchange, either between defects or between defects and impurities. A primary goal of research in this area is the identification and understanding of the physical properties of all the charge state dependent stable and low energy metastable states of a defect. With few exceptions, this is a sizeable undertaking because it requires the examination and analysis of an extensive set of structures for various charge states. The enormous computational effort required for this purpose has hindered a complete theoretical examination of the structural and electronic properties of the most fundamental point defects in tetrahedral semiconductors.

Reliance on the most advanced theoretical techniques for electronic and structural studies has the advantage of high accuracy and reliability of the results. These methods are generally quite complex, however, and by their very nature limit the number of options that can be tested within a reasonable time, even on the fastest computers currently available. Optimally, one would like to have a modelling approach that is both reliable and efficient. In this way many ideas can be quickly sorted through and the most promising ones retained for further study. In the work described below a combination of simple and usually reliable empirical tight-binding (Chadi 1978, 1979, 1984) and more sophisticated *ab initio* pseudopotential methods (Ihm *et al.* 1979) were used to optimize the search and characterization of native defects and impurities in GaAs.

A three-dimensionally periodic supercell each containing one impurity or point defect was used in all the calculations. It is quite remarkable that unit cells containing as few as 20–30 atoms can be used to decipher the properties of isolated point defects. The tight-binding energy-minimization procedure was generally used in the beginning phase of the calculations to improve an initial 'guess' geometry and for a quick evaluation of the merits of an idea before the start of self-consistent pseudopotential calculations. A tight-binding approach with only nearest-neighbour

Phil. Trans. R. Soc. Lond. A (1992) **341**, 247–253
Printed in Great Britain
© 1992 The Royal Society

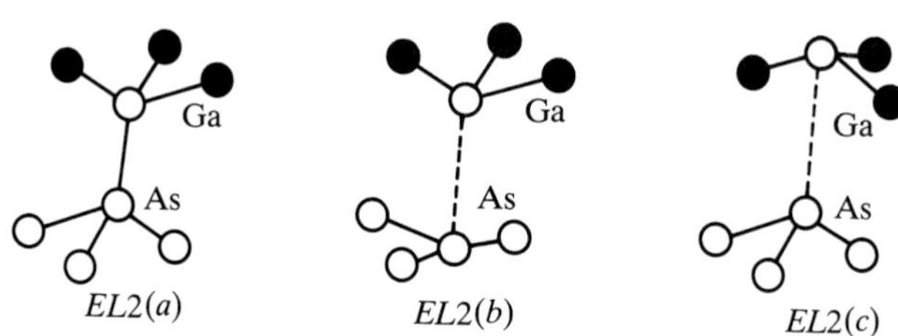

Figure 1. The EL2 (a), EL2 (b) and EL2 (c) configurations for an As-antisite in GaAs are shown. Dashed lines denote broken bonds. The EL2 (a) state is the stable state in semi-insulating and p-type GaAs. The EL2 (b) state is metastable in semi-insulating GaAs but is the stable state in n-type GaAs. The EL2 (c) state is an alternative low energy form of the antisite.

interactions has the great advantage that any reasonable bonding structure, starting with atomic coordinates far from their final equilibrium positions for that particular bonding topology can be examined. As a result, total-energy minima in the configuration coordinate diagram can be found more easily with the tight-binding method than with other methods. The method is also ideally suited for checking the convergence of *ab initio* total energies and atomic relaxations with respect to cell dimensions. This can be done by evaluating the magnitude of residual relaxations and energy changes for larger unit cell sizes, keeping the defect 'core' fixed.

2. The EL2 defect centre

The first defect problem to be examined by this combination of techniques was the As antisite defect in GaAs which results from the placement of an As atom on a Ga site. The defect is known to be at the core of an extensively studied centre labelled EL2 in GaAs (von Bardeleben *et al.* 1986). A most important property of EL2 is its light-induced metastability. The defect can be optically quenched by exposure to *ca.* 1.2 eV sub-band gap light, i.e. the optical absorption is initially strong but it decays with time and cannot be recovered unless the sample is heated to above 100 K in semi-insulating samples. It was originally believed that a simple substitutional defect such as the As-antisite had no low energy metastable states to account for this behaviour and a defect complex consisting of an antisite bound to an interstitial was proposed. An examination of the optical absorption spectrum (Kaminska *et al.* 1985) as a function of uniaxial pressure, however, gave impetus to the idea that EL2 was nothing more than the isolated antisite (labelled as EL2 (a) in figure 1). This result, which has received more support recently from photoluminescence measurements (Nissen *et al.* 1991), motivated a search for a possible metastable state of the antisite (Chadi & Chang 1988a; Dabrwoski & Scheffler 1988, 1989). The initial search was carried out using the empirical tight-binding method. No metastability was found for small arbitrary displacements of the As antisite from its substitutional site. A low energy state, labelled by EL2 (b) in figure 1, was found, however, for a displacement which ruptured an As–As bond. The results of the tight-binding calculations indicated an average charge transfer from the antisite to the other threefold coordinated As atom of about 0.75 e. Without inclusion of coulombic repulsion effects, the energy of the broken bond state was actually found to be *lower* than the fully bonded configuration. The identification of this low energy state of the antisite prompted us to do a full scale *ab initio* pseudopotential calculation which confirmed the large charge transfer and gave a more accurate estimate of 0.25–0.35 eV for the energy difference between the stable EL2 (a) and metastable EL2 (b) states (Chadi & Chang 1988a). The results of the calculations for an 18 atom cell were subsequently found to be in very good agreement with detailed studies on a larger 54 atom cell (Kaxiras & Pandey 1989).

More recently we have found a new metastable state of the defect, labelled EL2 (c) in figure 1. Tight-binding and self-consistent pseudopotential calculations indicate that the energies of the EL2 (b) and EL2 (c) configurations are within 0.1 eV $\pm$ 0.1 eV of each other. The EL2 (c) geometry has a lattice-relaxation-sensitive occupied state at 0.09 eV $\pm$ 0.1 eV above the VBM and an unoccupied state only a few hundredths of an electron volt above the conduction band minimum. The small energy difference between the EL2 (b) and EL2 (c) states is indicative of a soft phonon mode at an As antisite. The energy variation in going from the EL2 (b) to the EL2 (c) state is very small with no true minimum at the EL2 (c) configuration.

We have also obtained new results on the properties of As-antisites in n-type GaAs. The stable state of EL2 and of the EL2 (a) configuration of an As antisite in semi-insulating of p-type GaAs have no acceptor levels in the band gap. Each has a deep double donor level which contains two electrons in the neutral state. Despite this, As antisite derived EL2 centres have been identified via electron paramagnetic resonance (EPR) experiments in n-type GaAs (von Bardeleben *et al.* 1987). The experimental data suggest that the number of EPR active EL2 centers in n-type samples is several orders of magnitude large than in p-type or semi-insulating GaAs. The observation of an EPR signal is a strong, but so far neglected, indicator that the stable state of an antisite in n-type GaAs, unlike that in semi-insulating or p-type GaAs, cannot correspond to the EL2 (a) configuration. The results of our calculations lead to the prediction that the stable and metastable states of the antisite become interchanged in going from semi-insulating or p- to n-type GaAs. In the latter case, the negatively charged EL2 $(b)^-$ configuration is 0.12 eV lower in energy than the (EL2 $(a)^0$ + e), where e denotes a free electron in the conduction band and superscripts specify charge states.

The optical depth of the acceptor level of EL2 $(b)^-$ from the conduction band is calculated to be approximately 0.5 eV, about 0.38 eV larger than its thermal ionization energy. My theoretical results also indicate the existence of a *double acceptor* state for the EL2 (b) configuration. This has led Jia *et al.* (1992) to assign the acceptor states of EL2 (b) to the E1 and E2 defects in GaAs. The threefold coordination of the antisite is consistent with their EPR-derived results.

The possibility suggested by an analysis of optically detected electron nuclear double resonance (ENDOR) experiments (Meyer *et al.* 1987) that the stable state of EL2 in semi-insulating GaAs has C_{3v} rotational symmetry consistent with that expected for an antisite-interstitial complex was investigated in detail. A large number of structures were examined but no satisfactory solution consistent with this picture was obtained. The most severe problem with the complex model is its high energy under the constraint of threefold rotational symmetry. The lowest energy configuration of the complex has a split-interstitial structure with no (or at most C_2) symmetry) similar to that proposed by Delerue *et al.* (1987). Imposition of C_{3v} symmetry raises the total energy by nearly 2.5 eV.

3. Substitutional donors

We have extended the range of our initial investigations (Chadi & Chang 1988*b*, 1989; Zhang & Chadi 1990) on Si and S substitutional donors in GaAs and AlGaAs to Ge, Sn, Se, and Te impurities. It was found that, in each case, the most stable state of the negatively charged DX centre has a structure consistent with that determined previously, i.e. for column IV impurities DX formation involves the motion of the donor atom into a threefold structure (similar to that of the EL2 (*a*) state), whereas for column IV impurities, a displacement of a nearest-neighbour Ga or Al atom into a broken bond 'interstitial' position is required. An alternative mechanism in which DX centres result from a tetrahedrally symmetric 'breathing mode' distortion of the nearest neighbours of a donor was also examined. The negative-U properties of DX centres were found to be preserved for the new state. The energies of the tetrahedrally symmetric DX centres in GaAs are calculated to be within 0.05 eV of those for the threefold symmetric geometries of Sn, Se and Te donors, suggesting that these configurations could occur under suitable pressure or alloying conditions. The tetrahedrally symmetric model is found to be less important for Si and Ge donors.

A third type of negatively charged DX centre was found most recently for Sn donors. In this case the centre arises from the motion of one of the As nearest neighbours of the Sn impurity into an 'interstitial' configuration. This is an analogue of the EL2 configuration for As antisites in GaAs which has nearly the same energy as the 'regular' EL2 state. The new DX structure for Sn is calculated to have an energy within 0.08 eV of the 'traditional' DX centre in which the Sn atom undergoes a large displacement. The three types of DX centres can be distinguished, in principle, by their different optical ionization energies.

4. Interstitial defects

We have recently embarked on an extensive study of the properties of interstitial defects in GaAs and have obtained a number of new and sometimes surprising results. In the initial phase of this work Ga and As self-interstitials in GaAs (and Si interstitials in Si) were studied and a much larger set of structures than in the original pioneering work of Baraff & Schluter (1985) was examined.

An important new result is that Ga and As interstitials have both donor-like and acceptor-like atomic configurations. Arsenic, and gathium interstitials are predicted to self-compensate in undoped GaAs, giving rise to negative-U systems with $U = -0.7$ eV for As and -0.2 eV for Ga. These results are a consequence of the properties of an unusual type of $\langle 110 \rangle$ split-interstitial binding that gives very low energies for several charge states of these defects.

The interstitial binding geometries were determined for four different charge states. The most stable state of a Ga^{+2} interstitial is found for a tetrahedral interstitial site where it is surrounded by four Ga atoms. The four atoms relax outwards by about 0.12 Å†. The other inequivalent tetrahedral site is 0.14 eV higher in energy. The Ga–interstitial bond length to the neighbouring four As atoms is equal to the bulk Ga–As bond length in this case even though there are more distant neighbour relaxations. The tetrahedral geometry for the spin active Ga^{+2} interstitial is consistent with experimental data of Lee *et al.* (1988) for Ga interstitials in GaP.

† 1 Å = 10^{-10} m = 10^{-1} nm.

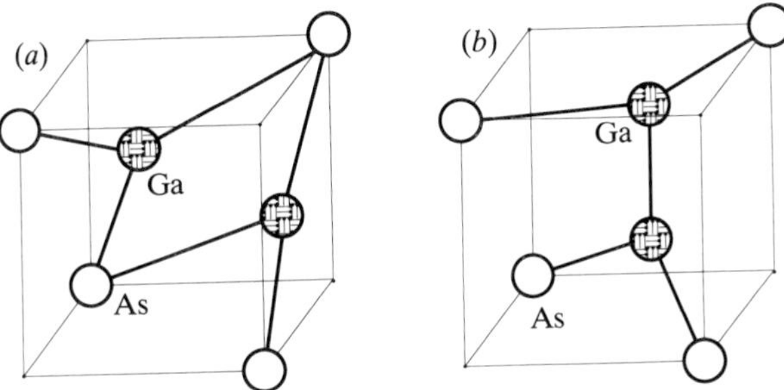

Figure 2. A new $\langle 110 \rangle$ split-interstitial geometry for a Ga interstitial defect in GaAs is shown in (*a*) and compared with the usual $\langle 100 \rangle$ split configuration in (*b*). The $\langle 110 \rangle$ split is unusual in that there is very little bonding between the split interstitials. A $\langle 110 \rangle$ split is found to have the lowest energy for several charge states of Ga and As self-interstitials in GaAs.

The singly charged Ga^+ interstitial has a totally different structure. Its lowest energy state occurs for a twofold coordinated configuration in which it breaks a bulk Ga–As bond and forms a bridge bond between the two atoms. This configuration is nearly 1.5 eV lower in energy than for a tetrahedral interstitial one. The neutral and negatively charged states have yet another structure: a $\langle 110 \rangle$ split-interstitial configuration shown in figure 2*a*. The bonding is quite distinct from the usual $\langle 100 \langle$ split shown in figure 2*b*. The Ga–Ga distance in the $\langle 110 \rangle$ split is 2.7 Å. As a result, the bonding between these atoms is extremely weak and essentially non-existent. The two split-interstitial Ga atoms are threefold coordinated and two of their surrounding four As atoms are fivefold coordinated. It is interesting to note that the $\langle 110 \rangle$ split gives an energy within 0.2 eV of the ground state for the *positively* charged twofold coordinated Ga interstitial.

Interchange of the Ga and As labels in figure 2*a* leads to one possible split-interstitial configuration for As. Among eight structures tested, this configuration has the lowest energy for the neutral and negatively charged states of As interstitials in GaAs. It is only for the $+2$ and $+1$ charged states of an As interstitial that a $\langle 100 \rangle$ like split-interstitial bonding geometry leads to the lowest energy. Charge transfer between the single donor and acceptorlike configurations is predicted to lower the energy of a pair of As interstitials by 0.7 eV. Such a large negative-U value for As interstitials in GaAs may explain why no magnetic resonance identification of any group V interstitial has ever been achieved in a III–V semiconductor, even under high energy electron irradiation conditions (Watkins 1992) where they are known to be produced in large numbers.

The results obtained for self-interstitials in GaAs were extended to Si interstitials in Si. Among all configurations examined, the new $\langle 110 \rangle$ split geometry is found to give the lowest energy for positive, neutral, and negatively charged states. For the doubly positive charge state, the $\langle 110 \rangle$ split is energetically nearly degenerate with the tetrahedral interstitial configuration. Binding at the hexagonal interstitial site gives a very low energy (about 0.1 eV) metastable state for both the positively charged and neutral states.

The work is now being further extended to B, Be and Al interstitials in Si. Preliminary results for Al indicate that the lowest energies for neutral and positively charged states occur for Al in a tetrahedral interstitial position. The negatively charged state is most stable in a twofold coordinated bridge geometry with the $\langle 110 \rangle$ split only slightly higher in energy. Experimentally, the double positive charge state of interstitial Al has been identified by EPR (Watkins 1964) and the defect resides in a tetrahedral site. There is also experimental evidence from diffusion studies (Troxell *et al.* 1979) for a substantial rebonding of Al interstitials in p-type Si upon minority carrier injection. This is consistent with our results on the change in bonding configuration when Al becomes negatively charged.

5. Conclusions

A brief review of our latest results on the structural metastabilities exhibited by substitutional and interstitial impurities and native defects in GaAs was given. Low energy metastable structures are found to be a general feature of these systems. Charge exchange either between defects or between defects and impurities plays a significant role in the energetics of defect reactions and has to be taken into account in any complete description of the properties of these centres. Arsenic antisites which are normally double donors in p-type and semi-insulating GaAs are predicted to become double acceptors in n-type GaAs. A new type of split-interstitial bonding that is important in the understanding of the ground state properties of self-interstitials and interstitial impurities in GaAs and Si was obtained. Arsenic interstitials in GaAs are found to be the largest negative-U defects discovered so far in a III–V semiconductor.

References

Baraff, G. A. 1989 *Phys. Rev.* B **40**, 1030.

Baraff, G. A. & Schluter, M. 1985 *Phys. Rev. Lett.* **55**, 1327.

Baraff, G. A. & Schluter, M. 1987 *Phys. Rev.* B **35**, 6154.

Chadi, D. J. 1978 *Phys. Rev. Lett.* **41**, 1062.

Chadi, D. J. 1979 *Phys. Rev.* B **19**, 2074.

Chadi, D. J. 1984 *Phys. Rev.* B **29**, 785.

Chadi, D. J. & Chang, K. J. 1988*a* *Phys. Rev. Lett.* **60**, 2187.

Chadi, D. J. & Chang, K. J. 1988*b* *Phys. Rev. Lett.* **61**, 873.

Chadi, D. J. & Chang, K. J. 1989 *Phys. Rev.* B **39**, 10063.

Dabrowski, J. & Scheffler, M. 1988 *Phys. Rev. Lett.* **60**, 2183.

Dabrowski, J. & Scheffler, M. 1989 *Phys. Rev.* B **40**, 10391.

Delerue, C., Lannoo, M., Stiévenard, D., von Bardeleben, H. J. & Bourgoin, J. C. 1987 *Phys. Rev. Lett.* **59**, 2875.

Ihm, J., Zunger, A. & Cohen, M. L. 1979 *J. Phys.* C **12**, 4401.

Jia, Y. Q., von Bardeleben, H. J., Stiévenard, D. & Delerue, C. 1992 In *Proc. 21st Int. Conf. on the Physics of Semiconductors.* (In the press.)

Kaminska, M., Skowronskii, M. & Kuszko, W. 1985 *Phys. Rev. Lett.* **55**, 2204.

Kaxiras, E. & Pandey, K. C. 1989 *Phys. Rev.* B **40**, 10044.

Lee, K. M. 1988 In *Defects in electronic materials* (ed. M. Stavola, S. J. Pearton & G. Davies), vol. 104, p. 449. Pittsburgh: Material Research Society.

Meyer, B. K., Hoffmann, D. M., Niklas, J. R. & Spaeth, J.-M. 1987 *Phys. Rev.* B **36**, 1332.

Mitonneau, A. & Mircea, A. 1979 *Solid State Commun.* **30**, 157.

Nissen, M. K., Villemaire, A. & Thewalt, M. L. W. 1991 *Phys. Rev. Lett.* **67**, 112.

Troxell, J. R., Chatterjee, A. P., Watkins, G. D. & Kimmerling, L. C. 1979 *Phys. Rev.* B **19**, 5336.

von Bardeleben, H. J., Stiévenard, D., Deresmes, D., Huber, A. & Bourgoin, J. C. 1986 *Phys. Rev.* B **34**, 72.

von Bardeleben, H. J., Miret, A., Lim, H. & Bourgoin, J. C. 1987 *J. Phys.* C **20**, 1353.

Watkins, G. D. 1964 *Radiation damage in semiconductors*, p. 97. Paris: Dunod.

Watkins, G. D. 1992 In *Materials science and technology* (R. W. Cahn, P. Haasen & E. J. Kramer) vol. 4. (*Electronic structure and properties of semiconductors* (ed. W. Schroter).) (In the press.)

Zhang, S. B. & Chadi, D. J. 1990 *Phys. Rev.* B **42**, 7174.

Discussion

A. M. Stoneham (*Harwell Laboratory, Didcot, U.K.*): There is a common belief that semiconductor interstitials have a large entropy. One apparent source of this could be an energy (formation or migration) that decreases rapidly as the cell volume increases with thermal expansion. Have you looked at how your predicted internal energies change with lattice parameter?

D. J. Chadi: The large lattice strains induced by an interstitial make it very likely that the total energy will decrease rapidly with increasing cell volume. I have not yet done any calculations on the lattice parameter dependence of the total energy but plan to do them in the future.

Modelling of catalysts and its relation to experimental problems

By C. Richard A. Catlow and John Meurig Thomas

Davy Faraday Research Laboratory, The Royal Institution of Great Britain, 21 Albemarle Street, London W1X 4BS, U.K.

We illustrate the role of both computer simulation and the evaluation of electronic structure in the study of solid heterogeneous catalysis by reference to recent work in this laboratory on (*a*) microporous materials (that have a spatially uniform distribution of accessible active sites) and (*b*) non-porous metal oxides. Computational methodologies may be used to model, first, the structure of the uniform catalysts both before and after thermal activation, second, the docking and diffusion of molecules in solids and on their surfaces; and, third, the reaction pathways of molecules at the active site. We highlight recent successes in modelling (i) the structures of zeolitic solids, (ii) the sorption of hydrocarbons within them, (iii) the protonation of small molecules at the Brønsted acid sites in uniform solid acid (zeolite) catalysts, and (iv) the reactions of small molecules on CeO_2 and MgO surfaces.

1. Introduction

Quantitative modelling is of greatest value when used directly in conjunction with experiment. It serves to confirm existing data, to interpret unusual or counter-intuitive discoveries, and to achieve aims not otherwise attainable except by synergistic use of computational techniques and experiment. The full potential of these methods has not yet been realized, but it is becoming increasingly possible to obtain information which is not in practice retrievable by experiment as well as to predict altogether new phenomena, structures and new processes.

Despite the considerable recent successes of quantitative modelling, there have not hitherto been many examples where it has been so indisputably ahead of experiment that new phenomena have been discovered through its sole deployment. Channelling of ion and atom beams in crystalline solids is one notable exception. More often, however, experiment has driven the modelling, in some instances for several decades. A good example of the latter is the 7×7 surface reconstruction of the silicon (111) surface which was discovered by low energy electron diffraction by Schlier & Farnsworth (1959). This reconstruction was more clearly identified experimentally by Takayanagi *et al.* (1985). It has very recently given rise to sophisticated *ab initio* total-energy calculations (Stich *et al.* 1992; Brommer *et al.* 1992), using parallel super-computers that 'predict' the details of the dimerization–adatom movements and stacking fault changes in the outermost layers of the silicon established by experiment. In reality, these sophisticated techniques do not really predict the 7×7 reconstruction: instead, they arrive at the same quantitative details that were first

Phil. Trans. R. Soc. Lond. A (1992) **341**, 255–268
Printed in Great Britain

© 1992 The Royal Society

derived by direct experiment. Another example, symptomatic of current trends in modelling of solids is the work of O'Keeffe *et al.* (1992) on predicted new low-energy forms of carbon. Prompted by the isolation of microscopic amounts of fullerenes (in particular C_{60}) – in this case 'prediction' is quite genuine – solids of hitherto unknown structures are shown to be of lower energy than well-known stable polymorphs. O'Keeffe *et al.* used a first-principles molecular-dynamics technique to find the relaxed atomic geometries and corresponding electronic structures of several simple novel forms of three-coordinated solid carbons. One of these forms, which they termed polybenzene, was found to have a substantially lower energy (0.23 eV per atom) than C_{60}. Its isolation has not yet been reported, but is the subject of current experimental work.

In the modelling of heterogeneous catalysts, great progress may be made on those that are *uniform*, in the sense that the active sites (see below) are distributed in a spatially uniform fashion throughout the solid, which has the extra advantage – because of the zeolitic structure – of having all atoms in the bulk accessible to the reactant species which diffuse within the micropores that permeate the entire crystalline solid. Owing to the translational symmetry of these structures of uniform heterogeneous catalysts, which are also monphasic, they are much more amenable to theoretical investigation than those catalysts that are multiphasic and multicomponent (Thomas 1988, 1990, 1992).

In this paper we concentrate on the application of modelling to zeolitic and metal oxide catalysts. As noted, the zeolitic, uniform heterogeneous catalysts are well suited to a variety of modelling techniques. Moreover, as we describe below, they have the extra advantage of having been the subject of extensive experimental study so that the task of blending modelling with direct experimental approaches is straightforward. Quantitative measurements have already been performed – in many instances by us and our collaborators – on a number of the properties and phenomena upon which we focus in this paper. These include the following.

1. The location, energetics and diffusivity of guest (potentially reactive) molecules inside a zeolitic catalyst or a realistic model analogue of such a catalyst (Freeman *et al.* 1991; Shubin *et al.* 1992).
2. The Brønsted acidity of the active site in zeolitic solid acid catalysts.
3. The migration of transition-metal ions, such as Ni^{2+}, that function as the locus of catalytic conversion in such processes as the cyclotrimerization of acetylene over (Ni^{2+}, Na^{+})-exchanged zeolite Y (Maddox *et al.* 1988; Couves *et al.* 1990; Thomas 1990).
4. The stability and dynamic properties of a range of germanium-containing zeolites (George *et al.* 1992*b*), where we are able to predict stability for a class of compounds that has not yet been prepared.

Our work on metal oxide catalysts is predominantly concerned with those (like many fluorite structured oxides, pyrochlores, perovskites, spinels and other relatively simple ionic-covalent oxide structures) that entail sacrificial loss of oxygen (Catlow *et al.* 1990). Experiments using isotopically labelled reactants or labelled solid catalysts reveal that structural oxygen is removed from the catalyst by the reactant, thereby rendering it, at least temporarily, a non-stoichiometric oxide. Gaseous oxygen (also a reactant) is then taken up by the catalyst, thus making good its anion deficiency. The cycle repeats itself for as long as the catalyst is active. As an example of the role of modelling in this field we describe the recent work of Sayle *et al.* (1992) which has shown how high-energy surfaces of CeO_2 may readily oxidize CO to CO_2.

We also highlight a recent study of Shluger *et al.* (1992) on the mechanism of catalytic dissociation of H_2 on the surface of MgO.

The central theme of this paper is that most of the key issues in contemporary studies of catalysis are amenable to investigation by theoretical techniques. They include the detailed *structure of the solid or surface* in or on which the reaction takes place and the way in which this is modified during the activation of the catalyst, the *nature and structure of the active site*, the *diffusion* towards and *docking* at the active site of sorbed molecules, and the *mechanisms of reaction* of the docked molecules. Moreover, as we shall show, computational techniques are able to identify the critical factors controlling catalytic performance and to guide the experimental modification of catalytic materials.

2. Methodologies

The complexity of the problems posed by catalytic systems necessitates the use of the widest possible range of computational and theoretical techniques, often in a concerted manner. Recent studies of ourselves and others have involved the full battery of *simulation techniques* (in which knowledge of electronic structure is subsumed into effective interatomic potentials). These include the following.

1. *Energy minimization*, in which the minimum energy configuration corresponding to the specified potential is obtained using an iterative numerical technique from a specified initial configuration. The method has now reached a high degree of precision in the modelling of both the crystal structures of complex inorganic materials (including microporous catalytic solids) and of surface structures: problems that are clearly of vital importance in the study of heterogeneous catalysis. Indeed, we show below that such calculations may be undertaken routinely on highly complex systems. In addition the methods may be used to simulate local structural changes around dopants and heteroatoms in solids and on surfaces and the positions and energetics of extra framework cations in microporous solids, examples of which will be given later in this paper. Another and related range of applications of minimization methods is to the location of the minimum energy site for adsorbed molecules in pores and on surfaces, and, as shown later, minimization methods may be fruitfully combined with other simulation techniques to allow such calculations to be undertaken in an automated manner.

2. *Monte Carlo*, in which ensemble averages are computed by numerical procedures which entail the generation of an ensemble of configurations by a series of random moves, with the probability of a configuration being accepted into the ensemble normally being dependent on its Boltzmann factor. The most obvious role of such methods, in the present context is in studying the distribution (and its temperature dependence) for adsorbed molecules over the available sites within a solid and on its surface, a good example of which is given in the work of Yashonath *et al.* (1988).

3. *Molecular dyamics*, in which kinetic energy is included explicitly in the simulations which solve numerically the classical equations of motion of the system simulated. The technique allows us to model the detailed dynamical behaviour of the system and to simulate diffusion directly. In the work described later in this paper classical microcononical ensemble (NVE) techniques have been used on an ensemble to which periodic boundary conditions have been applied to generate an infinite system.

The quality of the interatomic potentials used in such studies is obviously of crucial importance. For ionic and semi-ionic solids, Born model potentials based on formal or partial charges have been extensively and successfully used. Short-range interactions are modelled by two and three body analytical functions, and where possible polarization effects should be included via a shell model formalism. Extensive reviews are available in Harding & Stoneham (1988) and Catlow & Cormack (1987). Molecular mechanics force fields, in which the conceptual starting point is the covalently bonded network (in contrast to the Born model which perceives the solid as an assembly of interacting ions) are being increasingly used for modelling microporous solids. Such models, which require the specification of bond-bending, bond-stretching and torsional parameters, must be used in describing the internal force field of sorbed molecules.

For both types of potential the crucial issue is parametrization. Empirical methods, which require fitting unknown parameters to the properties of model compounds, have been widely used. The availability of increasingly high quality potential energy surfaces from quantum mechanical methods using both Hartree–Fock and local density techniques (Gale *et al.* 1992; Purton *et al.* 1992; Harrison & Leslie 1992) in both cluster and periodic boundary calculations is leading to a new level of precision which has greatly extended the predictive capacity of simulations.

4. *Electronic structure* methods may use semi-empirical or *ab initio* techniques employing single and multi-determinantal Hartree–Fock procedures or local density approximation (LDA) methods. And calculations may be performed on both cluster or periodic systems. All these approaches are reviewed elsewhere in these proceedings. In the context of catalytic studies we should emphasize the following factors.

(*a*) *Semi-empirical techniques* continue to have an important role. They allow us to study large numbers of atoms, and to obtain approximate information on geometries which may then be refined by *ab initio* calculations. These are particularly helpful in assessing properties such as the Brønsted acidity of a zeolitic catalyst (see below).

(*b*) Particular care must be paid to the *embedding techniques* used in cluster calculations. It is necessary that these are such as to reproduce accurately the electric potential and fields within the cluster and that the termination effects associated with the outermost atoms are minimal. Procedures discussed by Shluger *et al.* (1992) in this volume allow us to interface simulation with quantum cluster techniques in a way which achieves accurate and consistent embedding.

(*c*) Both LDA and Hartree–Fock methods have valuable roles to play. The former may allow larger clusters to be studied at the *ab initio* level, but the latter may profit from the substantial experience gained with their use in molecular quantum chemistry.

In summary we repeat the importance of a concerted strategy in theoretical studies of catalytic materials. Simulations are the appropriate technique for studying the structure of the solid and the docking and diffusion of sorbed molecules within it. Electronic structure methods allow us to model the reaction pathways of the molecule at the active site. Simulation and quantum mechanical methods may be combined as in the calculations discussed by Shluger *et al.* (1992); and, as already noted, electronic structure methods are playing a growing role in yielding high-quality potentials for subsequent use in simulations.

Table 1. *Energies* (kJ mol^{-1} TO_2 $unit^{-1}$) *relative to the quartz structure for* SiO_2 *and* GeO_2 *forms of various microporous structures*

	energy	
	SiO_2	GeO_2
silicalite I	11.2	12.5
mordenite[a]	20.5	16.1
faujasite[a]	21.4	24.5

[a] Whereas silicalite may be prepared (in either of two polymorphic forms I and II or in intergrowths of the two) in the pure siliceous state, neither faujasite nor mordenite may be synthesized in the pure siliceous form. Each of them can, however, be dealuminated to Si/Al ratios in excess of 100.

3. Recent applications

(*a*) *Modelling structures*

A high degree of accuracy is now possible in crystal structure modelling using minimization methods with currently available potentials. Several examples are provided by microporous solids and include the study of the small monoclinic distortion in silicalite by Bell *et al.* (1990) and the impressive recent study of Shannon *et al.* (1991) on the new zeolite Nu87: here modelling work revealed a low symmetry distortion which allowed a full refinement of high resolution power diffraction pattern for the material which had not been refineable using previously proposed higher symmetry structures.

The capability of modelling methods for structure *refinement* is therefore clearly established. More exciting applications concern the *prediction* of new structures. In this context the recent work of George *et al.* (1992*b*) is of particular note. This study examines the stability of a number of germanium analogues of zeolites. The stabilities of several GeO_2 'zeolitic' structures relative to those of the GeO_2 quartz structures are given in table 1 where they are compared with siliceous versions of the same structural types. Although there are significant differences between the relative energies of the SiO_2 and GeO_2 polymorphs, the results indicate that, in general, microporous Ge compounds should have roughly similar relative stabilities to the siliceous analogues. It would therefore be of considerable interest to attempt to synthesise these compounds. Already the synthesis and structure of a new microporous anionic derivative of GeO_2 has been accomplished in this laboratory (Jones *et al.* 1992).

Perhaps the most challenging problems concern the modelling of the structural changes during catalyst activation. A study of George *et al.* will show how detailed information may be obtained for a widely studied (Maddox *et al.* 1988) material, namely Ni-zeolite Y, for which activation is known to involve migration of Ni^{2+} ions from the hexagonal prism (S_I) sites into supercage positions as shown schematically in figure 1*a*. In a preliminary study George *et al.* (1992*a*) showed how the S_I, Ni^{2+} ion is stabilized by an extensive relaxation (0.5–0.8 Å†) of the surrounding oxygen ions: a result which nicely explains recent EXAFS data (of Dooryhee *et al.* 1990, 1991) on the dehydrated Ni-zeolite Y system. More detailed simulations were, however, able to follow the change in the energy of the Ni^{2+} as it moves out of the S_I site, through

† 1 Å = 10^{-10} m = 10^{-1} nm.

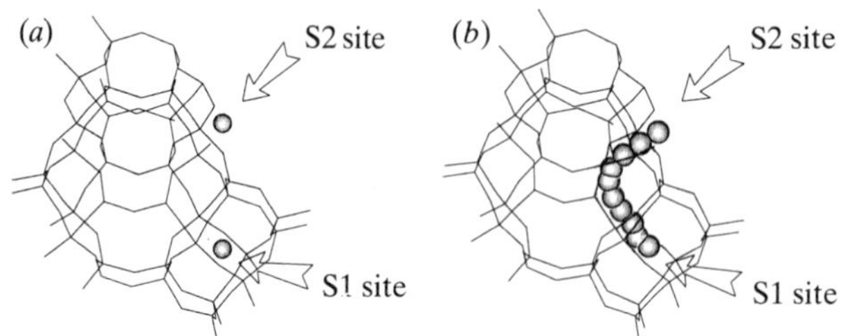

Figure 1. (a) S_I and S_{II} sites in zeolite Y. (b) Schematic trajectory of Ni ion migrating between S_I and supercage sites.

the sodalite cage and into the supercage, as illustrated in figure 1b. The highest energy point on this trajectory has an energy of *ca.* 2 eV relative to the S_I, Ni^{2+}, ion. Further calculations using *ab initio*, quantum mechanical techniques investigated the interaction of the Ni^{2+} ion at the saddle point with ligands including water NH_3 and acetylene situated at the neighbouring six-ring site. These molecules are known to assist (Couves *et al.* 1990) the activation of the catalyst; and the magnitude of the calculated interaction energies, which preliminary results suggest is in the range 2–3 eV indicates that such interactions are sufficient to overcome the energy barrier to motion from the S_I site supercage. The calculations have therefore allowed us both to follow the changes in ionic positions occurring during the activation of the catalyst and to understand how this is promoted by various reagents. We know from experiment, that reactant acetylene first 'draws out' the Ni^{2+} from the S_I site. Only after there are Ni^{2+} (or Ni^+) ions in the supercage does cyclotrimerization of the acetylene to benzene ensue.

(b) *Sorption, docking and diffusion*

(i) *Locating energy, minima*

The value of energy-minimization procedures in modelling the docking of molecules in zeolites was shown by the early work of Wright *et al.* (1985) who located the site occupied by pyridine in zeolite L. More recently Titiloye *et al.* (1990) reported a detailed series of energy minimization calculations on hydrocarbons in the C1–C8 range obtaining sorption energies in good agreement with experiment; these calculations also showed the important role of framework relaxation in lowering the sorption energy (*ca.* 20 kJ mol^{-1} for the larger hydrocarbons).

A substantial technical development was reported by Freeman *et al.* (1991), whose approach has been exploited and extended recently by Shubin *et al.* (1992). This novel method involves a combination of molecular dynamics (MD) Monte Carlo (MC) and energy minimization (EM) methods. The MD of the isolated molecules is undertaken in order to generate a library of low energy conformational states; each of these is then introduced randomly into the zeolite, and the resulting configuration is retained if its energy falls below a specified threshold value. All accepted configurations are then submitted to energy minimization. In the initial study of Freeman *et al.* (1991) the framework was constrained to be rigid, but in the more recent work of Shubin *et al.* (1992) framework relaxation was included.

Freeman *et al.* (1991) considered the topical case of the sorption of butene isomers in ZSM-5, which catalytically converts but-1-ene into isobutene (which is subsequently used in synthesis of the methyl tertiary butyl ether (MTBE) blending

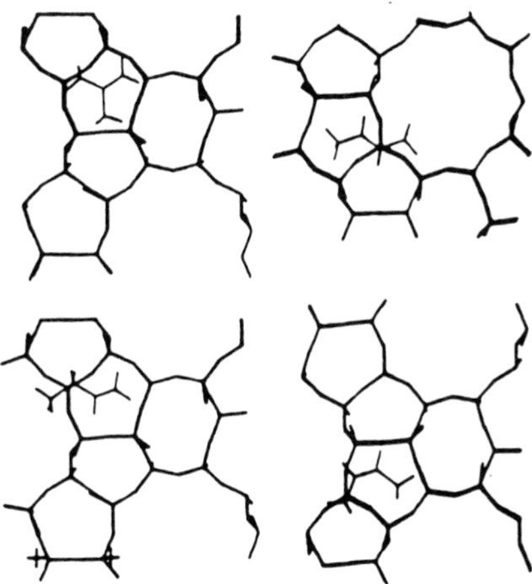

Figure 2. Low energy configurations for the four isobutene isomers in silicalite.

Table 2. *Butene isomer binding energies in silicalite*

isomer	calculated minimum energy/(kJ mol^{-1})	calculated average energy/(kJ mol^{-1})	relative minimum energy/(kJ mol^{-1})	relative average energy/(kJ mol^{-1})
2-methyl propene	−38.58	−28.09	0	0
but-1-end	−54.30	−44.41	−15.72	−16.30
cis-but-2-ene	−45.45	−31.64	−6.87	−3.55
trans-but-2-ene	−48.76	−38.44	−10.18	−10.35

agent). Their results for the four isomers are presented in table 2 which gives the minimum energy configurations (illustrated in figure 2) and the average energies of the docked minimized structures. By both criteria it is clear that isobutene is the least strongly adsorbed of the isomers: a result which is fully consistent with it being the dominant product in the isomerization. Moreover, the automated nature of these techniques would permit their use in a routine manner to screen modified or different zeolites to maximize the shape-selective discrimination between the different isomers.

The work of Shubin *et al.* (1992) explored the sorption of butanols in silicalite H-ZSM-5, following earlier experimental catalytic studies of Williams *et al.* (1990) and Stepanov *et al.* (1992). The results are summarized in table 3 while in figure 3 we display two of the energy minima. The calculations reveal that for all four butanol isomers, the predicted adsorption energies are only slightly different from one another. Second, the total adsorption energy increases only by 6–15 kJ mol^{-1} in H-ZSM-5 which is less than the expected value of 25–30 kJ mol^{-1}. This may be due to pore-confinement effects in the adsorption of C4 alcohols in H-ZSM-5. It seems that it is difficult to find any conformation for the C4 alcohol in the vicinity of the bridging OH-group in H-ZSM-5 without increasing the internal strain of the alcohol molecule and of the zeolite lattice. For *iso*-butanol, we can explain the similar adsorption energies in both silicalite and H-ZSM-5 as this is a bulky isomer while for both butanol-1 and butanol-2 and also for *tert*-butanol it is easier to optimize their positions for each within one of the zeolite channels. We also note, that, as a rule, for adsorption on both silicalite and H-ZSM-5, the energy values obtained for

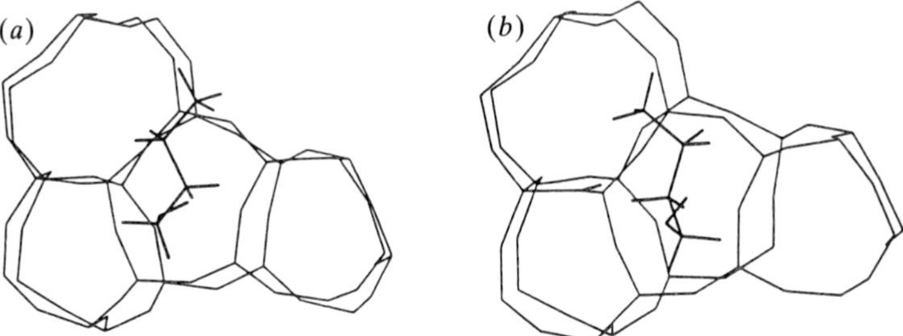

Figure 3. Examples of final butanol-1 positions in relaxed framework structure. (*a*) Butanol-1 in zig-zag channel of silicalite and (*b*) butanol-1 and bridging hydroxyl group (shown in bold lines) at position (19, 43, 4) in H-ZSM-5 [10].

Table 3*a*. *Low sorption sites and energies for butanol isomers in silicalite*

(All sorption energies are with respect to the zeolite and molecules at infinity and are given in kJ mol^{-1}. Channels: s, straight; si, straight near intersection; z, zig-zag; zi, zig-zag near intersection.)

butanol-1	butanol-2	*tert*-butanol	*iso*-butanol
−132.25 (z)	−117.47 (z)	−114.18 (z)	−121.31 (s)
−117.44 (z)	−111.69 (s)	−113.85 (s)	−119.87 (z)
−117.37 (s)	−102.67 (z)	−111.56 (z)	−119.79 (si)
−116.47 (si)	−102.27 (zi)	−109.54 (si)	−118.11 (si)
	−100.58 (zi)	−108.93 (zi)	−108.04 (zi)

Table 3*b*. *Low energy sorption sites and energies for butanol isomers on H-ZSM-5* (kJ mol^{-1})

(A cross (×) marks positions for which the lattice relaxation calculations were unsuccessful due to excessive distortions in the framework structure.)

crystallographic position of OH[a]						
Al	O	Si	butanol-1	butanol-2	*tert*-butanol	*iso*-butanol
2	13	8 (s)	−115.09	−112.93	−116.67	×
7	17	16 (z)	×	×	−129.06	−119.10
20	33	19 (si)	−119.52	−126.50	−124.32	−119.88
7	7	8 (si)	×	×	×	×
19	43	4 (z)	−138.42	−126.83	−120.37	−114.57
16	17	7 (z)	−120.42	−109.18	−105.09	×
14	32	18 (z)	−127.36	×	−110.98	−117.20
1	1	2 (s)	×	−113.08	×	×

[a]Using notation of Schröder *et al.* (1992*b*).

positions in straight and zig-zag channels and channel intersections are not very different. At all but the lowest temperatures, we would therefore expect the butanols to be distributed over a variety of sites.

Modelling methods may therefore predict sorption sites and energies both routinely and reliably. We now consider the more demanding problems posed by simulating dynamical behaviour.

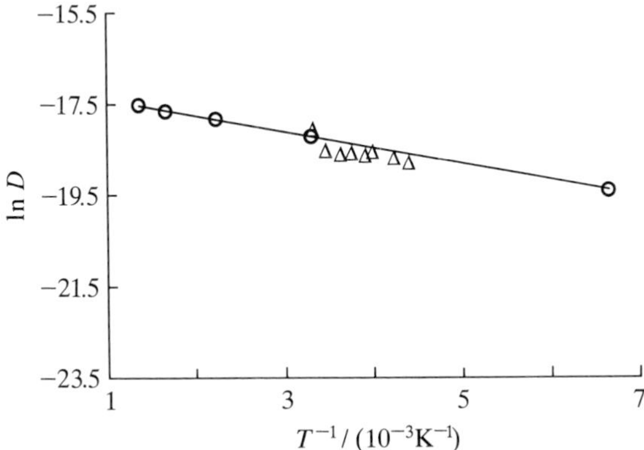

Figure 4. Calculated Arrhenius plot of CH_4 in silicalite obtained using MD simulations by Kawano *et al.* (1922).

Table 4. *Calculated diffusion coefficients* ($cm^2\ s^{-1}$) *for hydrocarbons in silicalite* (Numbers in parentheses indicate concentration in molecules per unit cell.)

molecule type	D (10^7)	molecule type	D (10^7)	molecule type	D (10^7)	molecule type	D (10^7)
methane (1)	410	methane (2)	1340	ethane (1)	692	ethane (2)	620
propane (1)	56.3	propane (2)	68	butane (1)	7	butane (2)	14
hexane (1)	14.4	hexane (2)	9.6				

(ii) *Molecular dynamics simulation of diffusion*

Recent studies (see, for example, Pickett *et al.* 1988; Demontis *et al.* 1988; Goodbody *et al.* 1991) have demonstrated the power of MD techniques in directly simulating diffusion of adsorbants in zeolites. The value of such studies in revealing both quantitative and qualitative information on diffusion rates and mechanism is shown by recent work of Kawano *et al.* (1992) and Hernandez *et al.* (1992*a*, *b*). These calculations, which included full framework flexibility were performed using standard microcanonical ensemble (NVE) MD. The results for methane are summarized in figure 4 in the form of an Arrhenius plot, which shows the excellent measure of agreement between calculated and measured diffusion coefficients.

Calculated diffusion coefficients for C1–C6 hydrocarbons are given in table 4. The values obtained for ethane and propane are in good agreement with experimental measurements obtained with the single-step frequency response (SSFR) (Den-Begin *et al.* 1989), and pulsed-field gradient (PFG) NMR techniques (Den-Begin *et al.* 1989; Caro *et al.* 1985). As noted by Den-Begin *et al.* (1989), values obtained with the PGF NMR techniques are approximately five times as large as those obtained with the SSFR method. Our values are compatible with both experimental measurements. The results for butane compare very well with those reported by Shen *et al.* (1991) measured with the SSFR technique, but those for hexane are higher than the reported experimental values of Den-Begin *et al.* (1989).

Analysis of the molecular trajectories shows that the molecules diffuse by means of 'jumps' whose distance is approximately equal to 0.5 lattice units (*ca.* 10 Å). For paraffins the channel intersections are energetically less favourable so that the molecules tend to reside in either the straight or the sinusoidal channels, and diffusion along a given channel will consist of a series of jumps across the channel

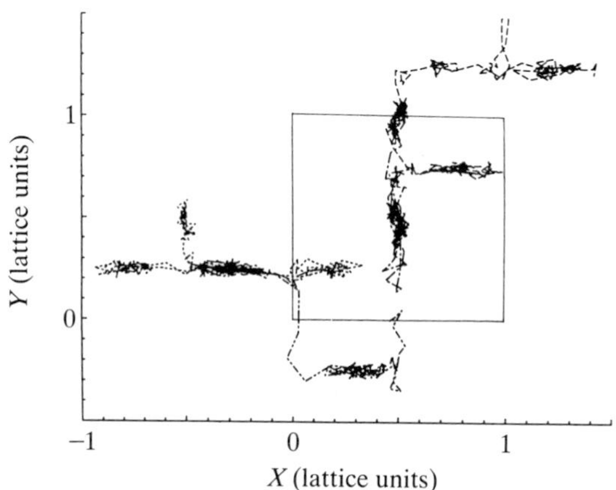

Figure 5. Trajectory plot for ethane in silicalite (at 300 K) illustrating movement of molecule between different channel systems.

intersections. However, a molecule that was originally diffusing along a straight channel can leave it through the sinusoidal channel. Examples of this kind of motion have been identified in our trajectory plots (as shown in figure 5).

A decrease in the diffusivity with increasing concentration of adsorbed molecules has been observed experimentally (Den-Begin *et al.* 1989; Caro *et al.* 1985) and predicted by simulation (June *et al.* 1990; Goodbody *et al.* 1991). As pointed out by Den-Begin *et al.* (1989), when the diffusion takes place by means of jumps, there are two possible mechanisms to explain this behaviour: (*a*) a reduction in the mean molecular jump length and/or (*b*) a decrease in the frequency of jumps. In our simulations we see no decrease in the mean jump length, though we cannot discount this type of mechanism at higher loadings.

The computational demands of such calculations should not be underestimated. It is necessary to run the simulations for long periods (> 100 ps) if reproduceable data are to be obtained. Such calculations are, however, becoming increasingly feasible with the new generation of parallel architectured machines which offer exciting opportunities for these types of investigation.

(*c*) *Modelling of active sites and reaction mechanisms*

In this, the most challenging of the fundamental problems posed by catalytic studies, electronic structure techniques clearly become necessary, although simulation methods still have a considerable role to play, as is illustrated by recent studies both by Brønsted acid sites in zeolites and of CO oxidation on CeO_2.

(i) *Brønsted acid sites and protonation reactions in zeolites*

A simplifying feature of zeolite catalysis is the fact that in many reactions the dominant active site is well defined, i.e. the protonated oxygen neighbouring a tetrahedral Al atom (or AlOH complex). Many studies using both quantum mechanical and simulation techniques have been reported of the fundamental properties of this centre (see, for example, Sauer 1992; Sim *et al.* 1991; Cheetham *et al.* 1984; Schröder *et al.* 1992*b*).

Recent studies of Schröder *et al.* (1992*a*, *b*) have demonstrated the value of simulation techniques in this field. Using lattice energy minimization methods based on effective potentials they investigated the Brønsted acid site in both faujasite and

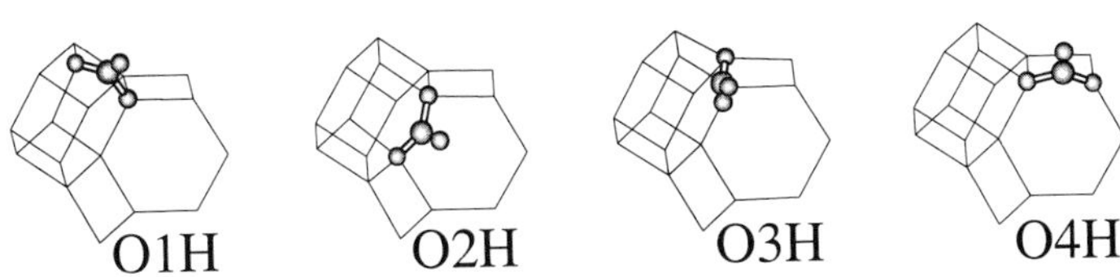

Figure 6. Brønsted acid sites in zeolite Y.

ZSM-5. The former is relatively simple as there is only one crystallographic T site, although there are four distinct types of oxygen site (see figure 6).

The calculations identify the O3 and O1 sites as being the most favoured for protonation (but with a small energy difference of *ca.* 5 kJ mol^{-1}). Calculations of the deprotonation energies indicate that the protonation of the O1 group results in slightly higher acidity. The calculations yield detailed information on the geometry of the AlOH group, showing, in particular, that the Al–O bond is surprisingly long (*ca.* 1.9 Å); they also yield vibrational frequencies that accord well with experiment. Moreover, the results from the simulation studies are in good agreement with earlier quantum mechanical calculation of Sauer *et al.* (1989). They establish the reliability of simulations (with carefully chosen potential parameters) in modelling the structures and properties of this type of acidic site.

The work of Schröder *et al.* (1992*b*) examined the more difficult problems of the location of the AlOH complex in ZSM-5, in which there are 12 crystallographically distinct T sites. They found that there are only small differences between the energies of the complex at different sites and, indeed, the results would lead us to expect a distribution that is close to random. These predictions contrast with those of previous studies (Derouane & Fripiat 1985); the latter work, however, was shown to be inadequate due to the omission of framework relaxation, which can be straightforwardly included in simulation studies.

Having identified the nature of the Brønsted acid site, the next stage is to study the mechanisms of the protonation reactions. Recent work has been undertaken in this field by J. D. Gale who has used both semi-empirical and *ab initio* techniques to investigate the protonation of small molecules (H_2O, NH_3, CH_3OH) at the Brønsted acid site in zeolites. The calculations used clusters which allows interactions of the molecule with both edge and face configurations to be studied. The results obtained on H_2O and NH_3 indicate the intriguing possibility of the protonated forms of these molecules docking above the face of the AlO_4 tetrahedron as illustrated for the case of NH_3 in figure 7. The stability of these configurations was supported by *ab initio* studies, although their energies with respect to the AlOH complex and the unprotonated molecules are still somewhat uncertain.

Further calculations are in progress aimed at examining the interaction of the Brønsted acid sites with organic molecules. The combination of the use of semi-empirical techniques for exploring a wide range of conformations with the use of *ab initio* techniques to refine the information on low energy structures is, we consider, a powerful strategy in catalytic studies.

(ii) *Oxide surfaces*

Again we find an important role for both simulation and quantum chemical methods in studying reactivity on oxide surfaces. As an example of the former, we cite the recent work of Sayle *et al.* (1992) who studied the interaction of CO with the surface of CeO_2. Simulations using an infinite two-dimensional periodic surface

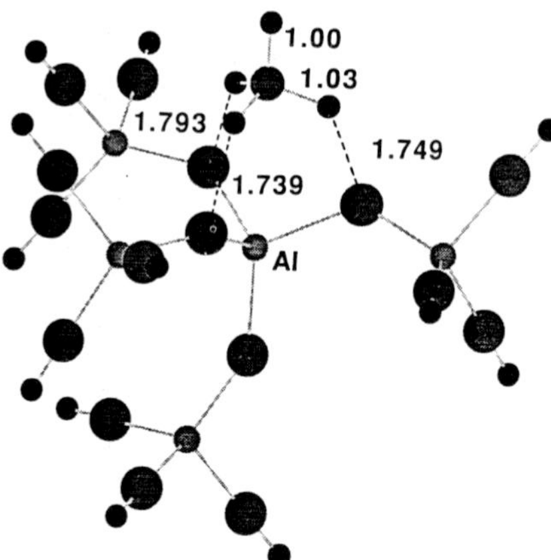

Figure 7. Protonated configuration of NH_3, above the face of an AlO_4 tetrahedron (as predicted by Gale (1992)). The protonated molecule binds to the tetrahedron via three hydrogen bonds.

(following the techniques developed by Tasker (1979) showed that both the (111) and (110) surfaces are stable with the former having the lower energy. Calculations of the energies to form surface vacancies were then performed using an adaptation due to Tasker (1979) of the general defect-simulation methodologies based on the procedure of Mott & Littleton (1938). The energies of formation of surface vacancies are found to be significantly, less in the case of the (110) surface. Moreover, using these calculated values (combined with standard electron affinity and bond energy data) we may estimate the energetics of the reaction:

$$O_s + CO \rightarrow V_s + CO_2,$$

where by O_s and V_s we indicate a surface oxygen atom and surface vacancy respectively. The values obtained are *ca.* 0.3 eV and -3.5 eV for the (111) and (110) surfaces respectively. The large negative value for the energy indicates that the surface could be highly active in promoting CO oxidation by release of oxygen from the solid in the classical sacrificial manner demonstrated recently for spinel catalysts in CO oxidation by Wright *et al.* (1992). Calculations of the barrier energies are in progress.

Our second example is provided by the recent study of Shluger *et al.* (1992) who investigated the dissociation of H_2 on the surface of MgO. Their work showed the efficacy of MgO molecules in effecting dissociation. Semi-empirical calculations on the interaction of H_2 with the MgO molecule 'adsorbed' on a large ($Mg_{25}O_{25}$) cluster indicated that the molecule dissociated to give an H–Mg–O–H species on the surface. A detailed examination using *ab initio* techniques of the dissociation mechanism for H_2 the reaction of H_2 with an isolated MgO molecule indicated a low barrier to dissociation of *ca.* 70 kJ mol^{-1}. The relevance of this type of efficient surface process in effecting catalytic dissociation is clearly demonstrated.

4. Conclusions

The recent results summarized in this paper illustrate how computational methods are able to guide the interpretation of experimental data in the field of catalytic studies; and moreover, that they are playing an increasingly predictive role in the field. With the availability of new techniques such as *ab initio* molecular dynamics and greatly enhanced computer power available in massively parallel architectures, predictive application in catalysis of theoretical and computational methodologies are certain to expand.

We are grateful to J. D. Gale, A. R. George, E. Hernandez, T. Sayle, A. Shubin and A. L. Shluger for their contributions to work discussed in this paper and for their permission to quote their unpublished results. We are grateful to SERC, The European Community, ICI plc and Johnson Matthey plc, for supporting this work.

References

Bell, R. G., Jackson, R. A. & Catlow, C. R. A. 1990 *J. chem. Soc. chem. Commun.* 782.

Brommer, K. D., Needels, M., Lanson, B. E. & Joannopoulos, J. D. 1992 *Phys. Rev. Lett.* **68**, 1355.

Caro, J., Bulöw, M., Schirmer, W., Kärger, J., Heink, W., Pfeifer, H. & Zdanov, S. P. 1985 *J. chem. Soc. Faraday Trans.* **81**, 2541.

Catlow, C. R. A. & Cormack, A. N. 1987 *Int. Rev. phys. Chem.* **6**, 227.

Catlow, C. R. A., Jackson, R. A. & Thomas, J. M. 1990 *J. phys. Chem.* **94**, 7889.

Catlow, C. R. A. & Price, G. D. 1990 *Nature, Lond.* **347**, 243.

Cheetham, A. K., Eddy, M. M. & Thomas, J. M. 1984 *J. chem. Soc. chem. Commun.* 1337.

Couves, J. W., Jones, R. H., Thomas, J. M. & Smith, B. J. 1990 *Angew. Chem. Int. Edn Adv. Mater.* **2**, 181.

Den-Begin van N., Rees, L. V. C., Caro, J. & Bülow, M. 1989 *Zeolites* **9**, 287.

Demontis, P., Suffritti, G. B., Quartieri, S., Frois, E. C. & Gamba, A. 1988 **153**, 551.

Derouane, E. G. & Fripiat, J. G. 1985 *Zeolites* **5**, 165.

Dooryhee, E., Greaves, G. N., Steel, A. T., Townsend, R. P., Carr, S. W., Thomas, J. M. & Catlow, C. R. A. 1990 *Faraday Discuss. chem. Soc.* **89**, 119.

Dooryhee, E., Catlow, C. R. A., Couves, J. N., Maddox, P. J., Thomas, J. M., Greaves, G. N., Steele, A. T. & Townsend, R. P. 1991 *J. phys. Chem.* **95**, 4514.

Freeman, C. M., Catlow, C. R. A., Thomas, J. M. & Brode, S. 1991 *Chem. Phys. Lett.* **186**, 137.

Gale, J. D., Catlow, C. R. A. & Mackrodt, W. C. 1992 *Model. Simulat. Mater. Sci. Engng* (In the press.)

George, A. R., Catlow, C. R. A. & Thomas, J. M. 1992*a* *Catalysis Lett.* **8**, 191.

George, A. R., Catlow, C. R. A. & Thomas, J. M. 1992*b* (Submitted.)

Goodbody, S. J., Watanabe, K., MacGowan, D., Walton, J. P. R. B. & Quirke, N. 1992 *J. chem. Soc. Faraday Trans.* **87**, 1951.

Harding, J. M. & Stoneham, A. M. 1986 *A. Rev. phys. Chem.* **37**, 53.

Harrison, N. M. & Leslie, M. 1992 *Molec. Simulat.* **9**, 171.

Hernandez, E., Kawano, M., Shubin, A. A., Freeman, C. M., Catlow, C. R. A., Thomas, J. M. & Zamaraev, K. I. 1992 In *Proc. Int. Zeolite Association Meeting.* (In the press.)

June, R. L., Bell, A. T. & Theodorou, D. N. 1990 *J. Phys. Chem.* **94**, 8232.

Jones, R. H., Chen, J., Thomas, J. M., Ruren, Xu. & George, A. R. 1992 *Chem. Mater.* **4**, 808.

Kawano, M., Vessal, B. & Catlow, C. R. A. 1992 *J. chem. Soc. chem. Commun.*, 879.

Maddox, P. J., Stachurski, J. & Thomas, J. M. 1988 *Catalysis Lett.* **1**, 191.

Mott, N. F. & Littleton, M. J. 1938 *Trans. Faraday Soc.* **34**, 485.

O'Keeffe, M., Adams, G. B. & Sankey, O. F. 1992 *Phys. Rev. Lett.* **68**, 2325.

Pickett, S. D., Nowak, A. K., Cheetham, A. K. & Thomas, J. M. 1988 *Molec. Simulations* **2**, 353.

Purton, J., Jones, R., Catlow, C. R. A. & Leslie, M. 1992 *J. Phys. Chem. Miner.* (In the press.)

Sauer, J. 1992 In *Modelling of structure and reactivity in zeolites* (ed. C. R. A. Catlow). London: Academic Press.

Sauer, J., Kölmel, C. M., Hill, J-R. & Ahlrichs, R. 1989 *Chem. Phys. Lett.* **164**, 193.

Sayle, T., Parker, S. C. & Catlow, C. R. A. 1992 *J. chem. Soc. chem. Commun.*, 977.

Schlier, R. A. & Farnsworth, H. E. 1959 *J. chem. Phys.* **30**, 917.

Schröder, K.-P., Sauer, J., Leslie, M., Catlow, C. R. A. & Thomas, J. M. 1992*a* *Chem. Phys. Lett.* **188**, 320.

Schröder, K.-P., Sauer, J., Leslie, M. & Catlow, C. R. A. 1992*b* *Zeolites* **12**, 20.

Shannon, M. D., Casci, J. L., Cox, P. A. & Andrews, S. J. 1991 *Nature, Lond.* **353**, 417.

Shen, D. M. & Rees, L. V. C. 1991 *Zeolites* **11**, 684.

Shluger, A. L., Gale, G. D. & Catlow, C. R. A. 1992 *J. phys. Chem.* (In the press.)

Shubin, A.-A., Catlow, C. R. A., Thomas, J. M. & Zamaraev, K. I. 1992 *Proc. R. Soc. Lond.* A (Submitted.)

Sim, F., Catlow, C. R. A., Dupuis, M. & Watts, J. D. 1991 *J. chem. Phys.* **95**, 4215.

Stephanov, A. G., Zamaralv, K. I. & Thomas, J. M. 1992 *Catalysis Lett.* **13**, 407.

Stich, I., Payne, M. C., King-Smith, R. D. & Lin, J-S. 1992 *Phys. Rev. Lett.* **68**, 1351.

Takayanagi, K., Tanishiro, Y., Takahashi, S. & Takahashi, M. 1985 *Surf. Sci.* **164**, 367.

Tasker, P. W. 1979 *J. Phys.* C **12**, 4977.

Thomas, J. M. 1988 *Angew. chem. Int. Edn* **27**, 1673.

Thomas, J. M. 1990 *Phil. Trans. R. Soc. Lond.* A **333**, 173.

Thomas, J. M. 1992 *Scient. Am.* **266**, 112.

Titiloye, J. O., Parker, S. C., Stone, F. S. & Catlow, C. R. A. 1991 *J. phys. Chem.* **95**, 4038.

Williams, C., Makarova, M. A., Malysheva, L. V., Parkshtis, E. A., Talsi, E. P., Thomas, J. M. & Zamaralv, K. I. 1991 *J. Catalysis* **127**, 377.

Wright, P. A., Thomas, J. M., Cheetham, A. K. & Nowak, A. K. 1985 *Nature, Lond.* **318**, 611.

Wright, P. A., Natarajan, S., Thomas, J. M. & Gai-Boyes, P. L. 1992 *Chem. Mater.* (In the press.)

Yashonath, S., Thomas, J. M., Nowak, A. K. & Cheetham, A. K. 1988 *Nature, Lond.* **331**, 601.

Discussion

A. M. Stoneham (*Harwell Laboratory, Didcot, U.K.*): Is it diffusion or the reaction step that is rate limiting? Can you model the rate of the reaction step yet?

C. R. A. Catlow: It depends on the system as to whether diffusion or reaction is rate limiting. We can model diffusion rates well; modelling of reaction rates is still a major challenge.

M. Lal (*Unilever Research, Port Sunlight Laboratory, U.K.*): To obtain reliable information on the long-time diffusion coefficients of adsorbed molecules, it is essential that the molecular dynamics computation is continued for times sufficiently long that the molecule is able to diffuse through a distance at least an order of magnitude greater than its own dimension. Do the present-day computers offer us a realistic prospect of calculating the diffusion coefficients in situations where activation energies associated with the diffusive process are reasonably high, say, 20 kT or more?

C. R. A. Catlow: You are correct that long simulation times are needed to obtain good diffusion coefficients in studies. However, with parallel architecture systems very long simulations are increasingly practicable.

The quantum-chemical basis of the catalytic reactivity of transition metals

By R. A. van Santen[1], M. C. Zonnevylle[2] and A. P. J. Jansen[1]

[1] *Schuit Institute of Catalysis, Laboratory of Inorganic Chemistry and Catalysis, Eindhoven University of Technology, 5600 MB Eindhoven, The Netherlands*

[2] *Koninklijk/Shell-Laboratorium, Amsterdam (Shell Research BV) P.O. Box 3003, 1003 AA Amsterdam, The Netherlands*

State of the art computational quantum-chemical methods enable the modelling of catalytically active sites with an accuracy of relevance to chemical predictability. This opens the possibility to predict reaction paths of elementary reaction steps on catalytically active surfaces. The results of such an approach are illustrated for a few dissociation and association reactions as they occur on transition metal surfaces. Examples to be given concern CO dissociation, carbon-carbon coupling and NH_3 oxidation. Reaction paths appear to be controlled by the principle of minimum surface atom sharing.

1. Introduction

Microscopically, a generic catalytic reaction consists of a self-regenerating reaction cycle built of elementary reaction steps. The catalytic reaction cycle starts with molecular adsorption and is followed by dissociation and recombination of adsorbed molecular fragments. The cycle is closed by desorption of the product molecules (Boudart & Djéga-Mariadassou 1984). The selectivity of a catalyst depends, among other factors, on the rules of dissociation and recombination.

We review the electronic factors that control chemisorption of molecules and atoms on transition-metal surfaces to use as a basis for an analysis of the rules that determine the reaction paths of dissociation and association.

A wide body of excellent reviews exist summarizing the experimental data on chemisorption phenomena (Ponec 1975; Somorjai 1981; Ertl 1983, 1990). Also many theoretical reviews are available (Hoffmann 1988; Shustorovich 1990; Norskov 1992; van Santen 1991). The transition-metal valence electrons of relevance to surface chemical bonding are contained in a relatively narrow d-valence band and a broad partly overlapping s, p-valence electron band. The number of d-valence electrons may vary to a maximum of ten valence electrons per atom. A tight-binding description, in which the metal one-electron wave functions are considered to be a linear combination of atomic orbitals, appears to mimic the electronic properties of the surface d-levels relevant to chemisorption reasonably well (Newns 1969; Grimley & Torrini 1973). For example, as expected, an orbital of an adsorbing molecule interacts to produce bonding as well as antibonding fragment orbitals (Hoffmann 1988; van Santen & Baerends 1991; van Santen & de Koster 1991). For transition-metal surfaces one observes experimentally that the interaction-energy with adsorbates decreases as the d-valence electron-band becomes more occupied. This is due to the gradual filling of antibonding surface fragment orbitals.

Phil Trans. R. Soc. Lond. A (1992) **341**, 269–282
© 1992 The Royal Society

A tight-binding description serves less well for the *s*-*p*-valence electrons. To a first approximation the electrons in these broad bands behave as free electrons. The electron concentration per atom is more or less constant across the transition-metal series and equal to approximately one electron per atom.

In the jellium model, which describes the metal as a free-electron gas (Lang & Williams 1978; Lundquist *et al.* 1979; Norskov & Lang 1980; Norskov 1982; March 1986) electron-transfer between metal surface and adsorbate is assisted by the screening of the resulting adsorbate charge by the surface image charge generated in the free-electron gas. Charge-transfer is especially important for atomic adsorbates, with their low unoccupied atomic orbitals. However, we will see that it is also very important for the dissociation of molecules.

Both molecular as well as atomic chemisorption energies change with metal as well as surface topology. However, the changes in interaction energy of chemisorbed molecules are usually significantly less than that of chemisorbed atoms. Also, the bond energy of admolecules to the surface is usually less than that of adatoms. This difference is essential for the dissociation of a diatomic molecule, because the energy required to break the molecular bond has to be recovered by the interaction of the resulting metal atoms with the metal surface. Only when the sum of the adatom energies exceeds that of the adsorbed molecule dissociation will become thermodynamically feasible (Benziger 1980; van Santen *et al.* 1990).

We discuss results obtained with semi-empirical methods (Hoffmann 1988; van Santen & de Koster 1991) as well as first principle quantum-chemical methods (Baerends & Ros 1973, 1975; Blomberg & Siegbahn 1983; Brandemark *et al.* 1984; Hermann *et al.* 1985; Panas & Siegbahn 1988; Panas *et al.* 1989; Yang & Whitten 1989, 1991). Results obtained by both methods lead to the same picture of surface reactivity. The main restriction of first principle approaches is the limited size of the metal clusters that can be studied at present. Because of size limitations semi-empirical calculations, which can treat larger clusters but are limited due to their dependence on parametrization, will continue to have to be used. For this reason results as obtained with the ASED-method (Andersen 1974, 1975), which is an extension of the extended Hückel method (Hoffmann 1963) are also presented. In §2 a short discussion will be given on the crucial point of cluster choices.

2. Molecular versus atomic adsorption

(*a*) *The dependence on d-valence electron occupation*

The generally high average energy of the LUMO (lowest unoccupied molecular orbital) of a molecular adsorbate as CO will usually result in the occupation of only bonding surface-fragment orbitals. As the d-valence electron occupation increases, this interaction will increase. The interaction with the LUMO is often called electron-backdonation, because admixture of LUMO with surface orbitals results in a partial occupation of the LUMO which is naturally empty before adsorption.

The higher energy of the LUMO of a molecule in comparison to that of its constituent atoms, implies that the energy for electron transfer to the molecule is less favourable than in the case of atoms. For example, for CO coordinated C-end down to a Co cluster (van Santen & Zonnevylle 1992; Zonnevylle & van Santen 1992) using the first principle Amsterdam local density approximation (LDA) code (Baerends *et al.* 1973; Baerends & Ros 1973, 1975), figure 2*a* shows the increasing fractional orbital occupation of the $2\pi^*$ LUMO of CO and the total overlap population between $2\pi^*$ and

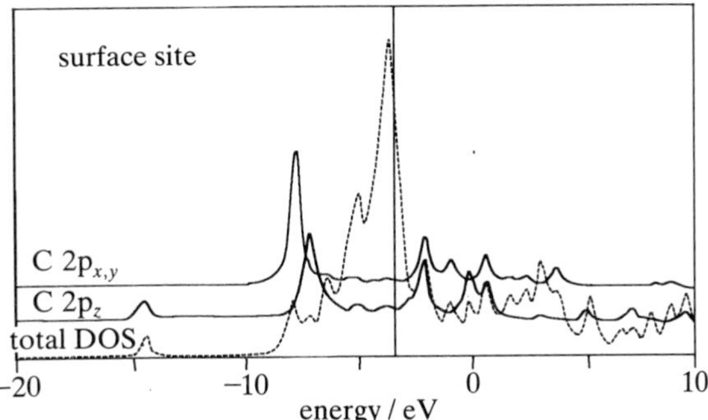

Figure 1. Local density of state C 2p$_x$, 2p$_y$ and 2p$_z$ atomic orbitals of C adsorbed to a nine atom Co cluster (z direction is surface normal). C is three fold coordinated 0.1 nm from surface plane. Results are due to a local density approximation calculation (Zonnevylle *et al.* 1990). Note the electron densities around 5 eV below the Fermi level that correspond to the maxima of the bonding adatom-surface electron-density distribution. The maxima corresponding to the antibonding electron-density distributions are approximately 2 eV above the Fermi level.

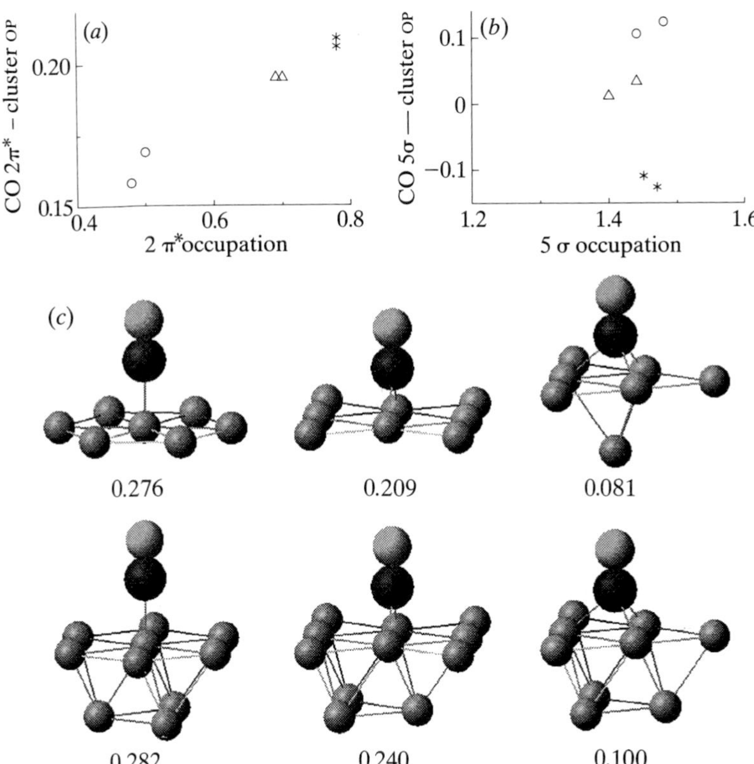

Figure 2. (a) 2π*-orbital population (OP) and 2π*-surface orbital overlap population of CO adsorbed to a Co cluster as a function coordination. The cluster simulates a (111) surface (van Santen & Zonnevylle 1992), see also (c). (b) 5σ-orbital population and 5σ-surface orbital overlap population of CO adsorbed to a Co cluster as function of coordination. The cluster simulates a (111) surface (van Santen & Zonnevylle 1992), see also (c). In (a) and (b): ○, top; △, bridge; *, hollow. (c) Clusters and geometries used in (a), (b) and figure 3. Within brackets total C–Co orbital overlap population (van Santen & Zonnevylle 1992).

all of the cluster orbitals. The fractional $2\pi^*$ orbital occupation reaches a value of close to 0.5 in the hollow high-coordination site. The relatively small values of the fractional occupation-number are indicative for the bonding nature of the $2\pi^*$-surface interaction, i.e. filling primarily adsorbate-surface bonding levels. This also follows from its increases with higher coordination number to the surface and the corresponding increase in $2\pi^*$-surface cluster orbital overlap population.

As illustrated in figure 1 in contrast to the molecular case, because of the low value of the average energy of the screened atomic LUMOs (a carbon atom in this case) with respect to the cobalt metal Fermi level now adatom–metal bonding as well as antibonding levels are occupied. Again in contrast to the molecular case an increase in d-valence electron occupation will increase the occupation of antibonding d-orbitals thus decreasing the interaction energy. However, the lower energy of the screened atomic orbitals compared with the molecular LUMOs contributes to the fact that adsorbate–surface interaction energies are greater than for molecules.

For a molecule, the interaction of its highest occupied molecular orbitals (HOMO), which are nearly always bonding combinations of intramolecular atomic orbitals, with the surface orbitals, is also very important. This can be seen most readily by analysing the bond-order overlap populations, which measure the contribution of each level of the adsorbate–surface complex with respect to its (positive or negative) contribution to the chemisorptive bond. One such bond-order overlap population curve is shown in figure 3 for the interaction of the CO 5σ HOMO orbital with the clusters. Note that this interaction specifically leads to occupied bonding as well as antibonding surface–adsorbate orbital fragments. The population of antibonding orbitals increases with coordination number. The total orbital overlap populations are presented in figure 2*b*.

The total CO 5σ fractional orbital occupation decreases with coordination number, corresponding to an adsorbate-to-substrate donative interaction. The donative interaction will usually be weak or, when the antibonding orbitals are nearly completely occupied, may even become repulsive (hollow coordination). An increase in the d-valence electron occupation would also enhance the repulsive nature of the donative interaction. For a detailed explanation of these chemical bonding features we refer to (van Santen & de Koster 1991; van Santen *et al.* 1990; van Santen & Zonnevylle 1992; Zonnevylle & Van Santen 1992).

Several important effects resulting from the choice of clusters are illustrated in figure 2*c* and table 1*a*–*c*. The interaction energies given in table 1*a* show significant changes with CO site for the 7–11 Co atom clusters. The variations between clusters in the total interaction energy are mainly due to differences in the indirect response of the clusters to the disturbance caused by the interacting CO molecule, rather than to the formation of the chemisorptive bond directly (table 1*b*). This effect can be ascribed to the differences in the net decreases in metal–metal bond strengths upon CO adsorption (table 1*c*). In contrast to the total interaction energy the overlap populations between CO and clusters are less influenced by cluster choice. The orbital overlap populations of the C–Co interaction (figure 2*c*) all indicate a stronger total interaction in atop versus bridging or hollow coordination.

As an overall result, the interaction of an adsorbed molecule changes much less with d-valence electron occupation than that of adsorbed atoms. The relative small variation of the molecule–surface interaction results from the fact that when the *d*-valence electron occupation increases, the increasing backdonative interaction is counteracted by a simultaneous decreasing donative interaction.

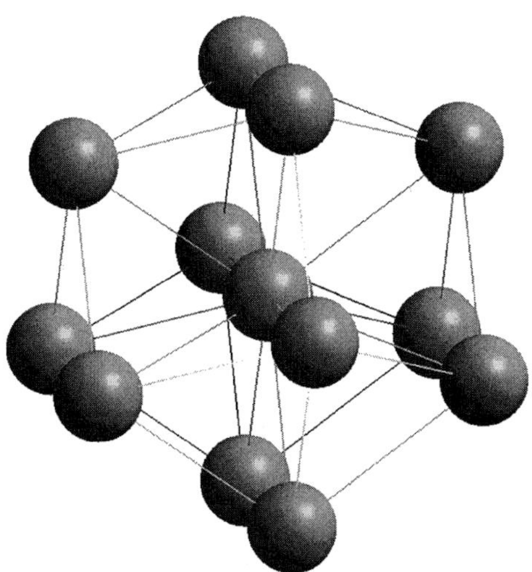

Table 1. (*a*) *Comparison of* CO *interaction energies* (kJ mol^{-1}) *as a function of* CO *coordination number computed for clusters 2c, with* CO *interaction energies computed for adsorption to the 13* Co *atom clusters* (*van Santen & Zonnevylle 1992*)

cluster size	13	9–11	7–8
onefold	−160	−200	−127
twofold	−140	−263	−218
threefold	−120	−241	−210

Table 1. (*b*) *Metal–metal atom bond weakening in the differently sized clusters, measured as percent change in bond order overlap*

(The situation of one-fold coordination is compared (van Santen & Zonnevylle 1992).)

cluster size	13	10	7
bond order overlap population change	−60	−32	−11

Table 1. (*c*) *Average* Co *metal–metal atom bond energy for clusters considered in* (*b*), *without* CO *adsorbed* (kJ mol^{-1}) (*van Santen & Zonnevylle 1992*)

bulk	13	10	7
71	100	111	140

For atoms such as C and O, however, there is only a significant change in the interaction with 2p atomic valence orbitals. The doubly occupied 2s atomic orbitals have a very low energy and their interaction is always repulsive). When the d-valence electron occupation increases, the interaction with the 2p atomic orbitals decreases because of the increased occupation of antibonding adsorbate–surface fragment orbitals.

The difference between molecular and atomic adsorption phenomenon is illustrated by the ASED calculation results presented in figure 4 (van Santen *et al.* 1990). The reaction paths illustrated in figure 4 are explained in the next section. By varying the

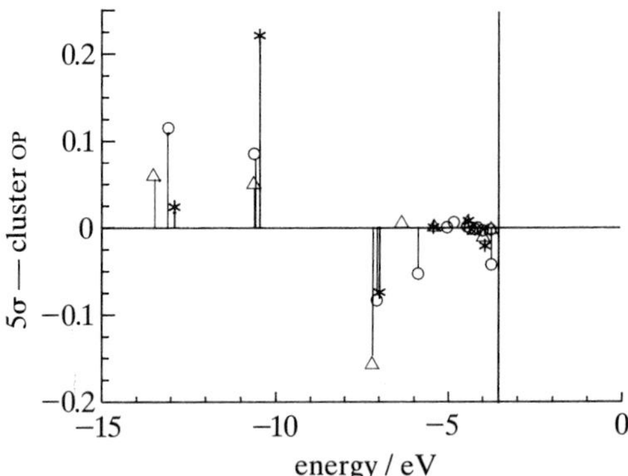

Figure 3. CO 5σ-surface orbital population as a function of energy on the same clusters as used in figure 2 as a function of position. The orbital population is positive when an orbital fragment is bonding and negative for the antibonding case. Note the minimum of energy of −12.5 eV of the bonding interaction contribution (van Santen & Zonnevylle 1992). *, Top 1; ○, bridge 1; △, hollow 1.

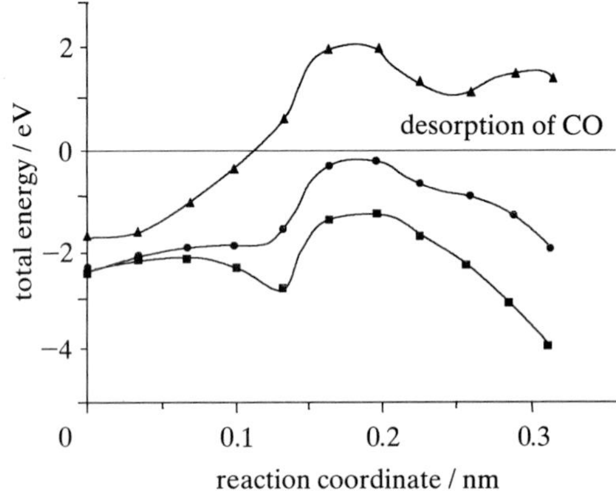

Figure 4. Change in ASED interaction energy of a dissociating CO molecule on the Rh (111) surface (19). The reaction coordinate is the projection of the CO distance on the surface. The calculations have been done for the minimum reaction energy path, to be discussed in §3*b* as determined for Rh (de Koster & van Santen 1990). ▲ denotes result for the same cluster with 1 electron per metal added (Pd). ● denotes result for the cluster with 1 electron per metal subtracted (Pd).

d-valence electron count (keeping the atomic parameters fixed), this semi-empirical method is well suited for tracing the effects of moving across the transition-metal series. As the d-electron occupation increases, the CO adsorption energy drops much less significantly than the C and O adsorption energies do. Although back-donative (M:CO) and donative (CO:M) effects change dramatically across the series, for the molecular adsorbate, the net effect remains in balance. Atomic adsorbates experience only an increasing donative, and thus increasingly repulsive, effect across the series.

(*b*) *The role of orbital symmetry*

The interaction of the adsorbed atom or molecule with the s, p-valence electron band usually dominates over that with the d-valence electrons. This is because the metal s, p-atomic orbitals are spatially more extended than the d-atomic orbitals. However, their interaction is more or less constant across the transition-metal series. The spatial extension of the d-atomic orbitals tends to increase from the right to the

Table 2. *Overlap population of the* $NH_3\sigma$ *orbital with selected copper orbitals (in arbitrary units) (Biemol et al. 1992)*

copper orbital	Cu(9,4,5)	Cu(8,6,2)
3d	−0.005	0.157
4s	0.693	0.372
4p	0.572	0.727
total	1.260	0.942

left across the transition-metal series as well as when moving downward along a column in the periodic system. This reflects in an increasing importance of the interaction with the d-valence electrons. For either molecular or atomic adsorbates, the grouping of adsorbate orbitals as either symmetric (s) or asymmetric (p) with respect to the surface normal, which we will choose along the z direction, aids the analysis.

When adsorbed to a single surface atom, by far the dominant interaction is with the metal atomic orbitals of the same symmetry. When adsorbed in higher coordination, one has to construct group-orbitals (van Santen 1987; Zonnevylle *et al.* 1989) for the surface atoms, i.e. linear combinations of nearest-neighbour metal-atomic orbitals, with coefficients given by the irreducible representation corresponding to the local symmetry of the adsorption site.

Whereas the p_x, p_y ad-atom orbitals cannot interact with the metal-atom s-atomic orbitals when adsorbed to a single surface-atom, in high coordination sites a strong interaction becomes possible with the corresponding s-atomic group-orbitals of p symmetry. Because the interaction with the s-valence electrons contributes significantly to the chemical bond strength, this causes adatoms to favour high coordination sites.

For molecules the same holds for the interaction with p-symmetric molecular orbitals, which are usually the LUMOs. So the interaction with LUMOs usually favours high coordination sites. This result was illustrated for the CO $2\pi^*$ orbital interaction in figure 2*a*. The HOMOs of the adsorbed molecule are often s-symmetric with respect to the surface normal. The relative energy of the screened HOMO-energy with respect to the s-valence-electron bond will determine, whether high or low coordination is favoured. For instance, the CH_3 fragment, with its main bonding contribution due to the σ-lone pair interaction (and little contribution from the LUMO) is found in high coordination sites on the Ni(111) surface (Schule *et al.* 1988). Ammonia, also mainly binding by way of its σ-lone pair orbital, will also favour high coordination on Ni (32). However, when interacting with a metal with a lower workfunction such as Cu, it favours low coordination sites. Whereas the interaction with the Cu 4p atomic orbitals increases with coordination number, it decreases for interaction with the Cu 4s as well as 3d atomic orbitals (see table 2).

As is shown in figure 5, for clusters modelling the (100) surface and using the LDA method, the low energy of the ammonia σ orbital results in a significant population of antibonding orbitals. For a full analysis of this feature in terms of grouporbital interactions we refer to (van Santen & Baerends 1991).

A repulsive interaction between the doubly occupied σ-NH_3 orbital and the doubly occupied d-valence electrons exists (see table 2). This repulsive interaction is proportional to the number of coordinating atoms and hence is less for low coordination. Generally one finds that the s-symmetric interaction of adsorbate

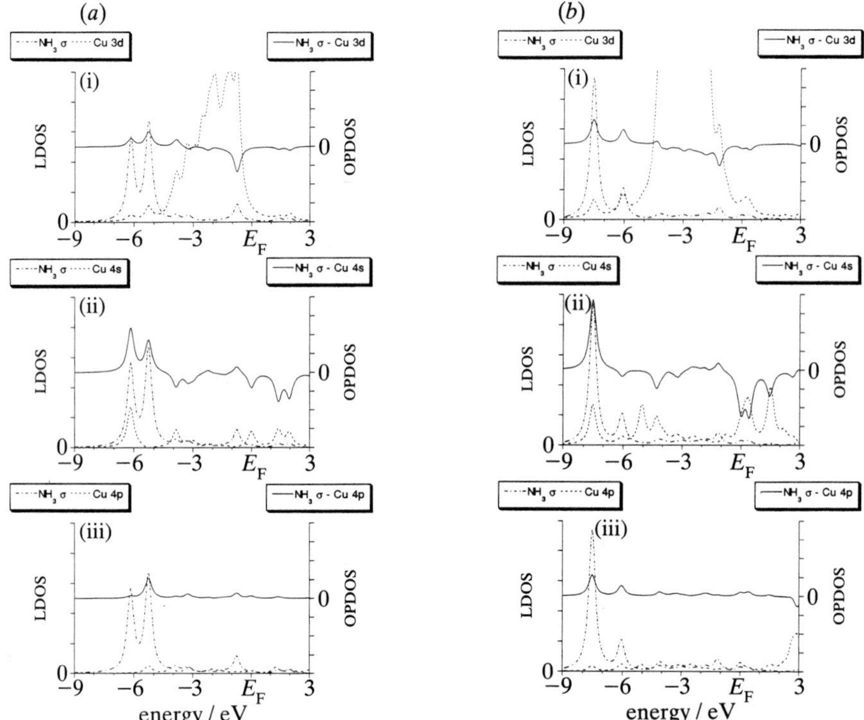

Figure 5. (*a*) One-fold coordinated NH_3 bonded to a (100)-surface type cluster. Local density of states (LDOS) of the NH_3 σ-orbital before adsorption, and the central copper 3d (i), 4s (ii) and 4p (iii) orbitals together with the overlap population density of states (OPDOS) between both the (100) surface cluster and NH_3. Zero energy in these figures corresponds to the Fermi level (E_F) (Biemolt *et al.* 1992). (*b*) Two-fold coordinated NH_3 bonded to a (100)-surface type cluster (Biemolt *et al.* 1992). LDOS of the NH_3 σ-orbital before adsorption, and the central copper 3s (i), 4s (ii) and 4p (iii) orbitals together with the OPDOS between both the (100) surface cluster and NH_3. Zero energy in these figures corresponds to the Fermi level E_F. LDOS and OPDOS are both in arbitrary units.

HOMO with the highly occupied d-valence electron band directs adsorbate molecules to low coordination sites.

The CH_3 fragment adsorbed to the Pt-surface has been predicted to adsorb atop of a Pt atom, due to the strong interaction with the spatially extended Pt 5d-atomic orbitals (van Santen *et al.* 1990).

The adsorption energy of a molecule is controlled by the balance of donating and back-donating terms. Back donation is relatively unimportant for CH_3 and NH_3 because of the high energy of their LUMOs. Back donation involving the LUMOs usually favours high coordination, donation involving the HOMOs can favour low coordination. The interaction with the d-valence-electron interaction controls the balance.

3. Dissociation and association reactions

(*a*) *The role of backdonation*

For dissociation, a molecular bond has to stretch. Assuming the molecular fragments thus generated remain adsorbed to the surface, the optimum configuration for the dissociating bond is parallel to the surface.

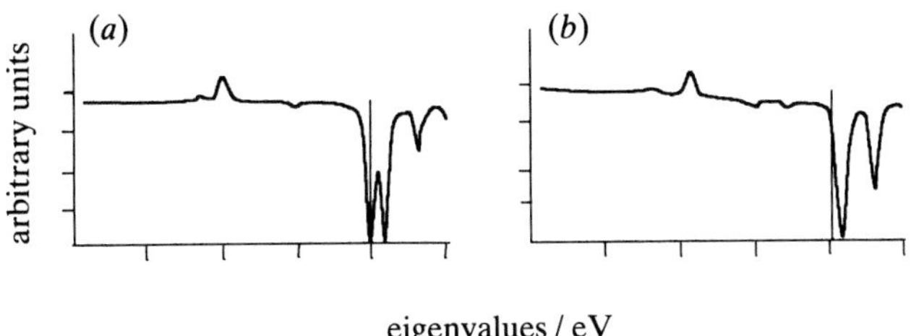

Figure 6. The changes in O–Ag overlap population of O adsorbed to a Ag_4 cluster, with O fourfold coordinated (Ag_4O) (*a*). With four additional oxygen atoms coordinated (*b*). Note the shift downwards of the antibonding AgO electron density when the oxygen concentration increases (van den Hoek *et al.* 1989).

Some molecules such as O_2 usually adsorb parallel to the metal surface (Upton *et al.* 1988; Kiskinova 1992), whereas others such as CO and NO (Kiskinova 1992) usually prefer perpendical adsorption with the carbon or nitrogen atom directed towards the surface. A molecule initially perpendicularly adsorbed has to bend before it will dissociate.

The back donating interaction between metal surface and p-symmetric antibonding LUMOs lowers the energy required of bond stretching. We have seen in §2 that the molecular LUMOs get a finite electron density. This causes weakening of the intramolecular bond, because of the antibonding nature of the molecule LUMOs with respect to the intra-molecular bond. For a diatomic molecule, adsorption parallel to the surface enables contact of two molecular atoms with the metal surface, which will favour the backdonating interaction with a LUMO, given surface group orbitals of proper symmetry are available.

However, when the d-valence electron occupation is high, interaction between the doubly occupied HOMOs of the molecule and the d-valence electrons is mainly repulsive. This tends to favour low coordination and minimal contact with the surface. It may even be the cause which forces diatomic molecules to adopt a perpendicular orientation with respect to the surface plane (Sanchez Marcos *et al.* 1990) in some instance.

If coordination is parallel to the surface the HOMO repulsive interaction with metal d-atomic orbitals may become overcome by shifting to a twofold coordination site where rehybridization of molecular HOMOs and LUMOs can occur (Sautet & Paul 1991; Chan *et al.* 1992).

Back donation is very strong on surfaces with low workfunctions and when a molecule has a high electron affinity. Then parallel adsorption may become the minimum energy configuration and high coordination sites are preferred.

(*b*) *Surface reaction paths*

Reaction paths that favour electron population of LUMOs of the dissociating molecule, as well as the development of a strong bonding interaction between the product fragments and the surface will tend to have a lower activation energy. We have already discussed that atoms and molecular fragments usually prefer bonding in high coordination sites. In addition, the product fragments should not be destabilized because of an unfavourable location with respect to each other on the metal surface. Such a destabilizing effect occurs when adatoms share a surface atom for bonding (see figure 6). So for the dissociation reaction an ensemble of accessible surface metal atoms is required.

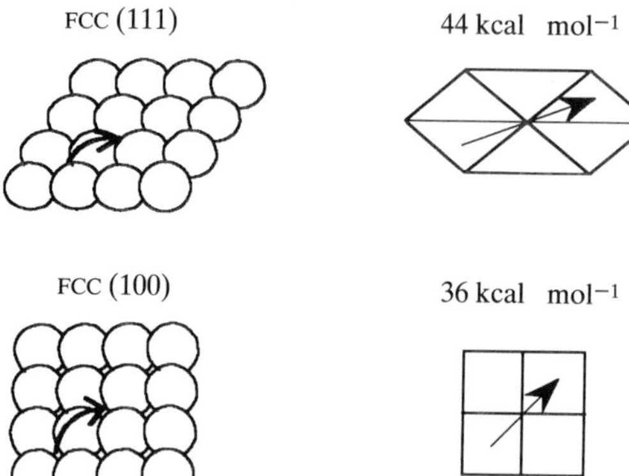

Figure 7. Minimum dissociation energy paths of CO (13) on large clusters of Rh according to the ASED method. The activation energy on the (111) surface is predicted to be 8 kcal mol^{-1} higher than on the (100) surface (de Koster & van Santen 1990). (1 cal ≈ 4.184 J.)

Several factors may contribute to such unfavourable interactions. On purely electrostatic grounds the direct interaction of negatively charged adatoms is unfavourable. Also, any metal adatom bond will weaken when that surface metal atom must also bind a second adatom. The effect of the latter stems primarily from the more restricted role that the s-surface metal atomic orbitals can play. The increased coordination of the metal atom with other atoms weakens the bonding interaction of the s-atomic orbital with each of the individual atoms. In essence, s-atomic electron density becomes localized in the chemical bond with the adatom. Hence less *s*-atomic electron density becomes available for other chemical bonds. For identical covalent bonds, the attractive contribution to the bond strength decreases approximately as $N^{-\frac{1}{2}}$ per bond (Cyrot-Lackman *et al.* 1974; Haydock *et al.* 1975; van Santen & Baerends 1991). (N is the number of next nearest-neighbour atoms of the central atom.)

This effect can be viewed as a consequence of the principle of bond order conservation (Shustorovich 1990), which gives as a general rule that the larger the number of bonds to a particular atom the weaker the strength per bond. Hence adatoms tend to share the least number of surface atoms as the surface coverage increases (Kiskinova 1991). It appears that the reaction paths for dissociation and association we have studied so far by theoretical means all seem to satisfy this principle of least sharing of surface-metal atoms.

Figure 7 illustrate the minimum energy paths found for CO dissociation (de Koster & van Santen 1990) on Rh (111) and Rh (100) according to the ASED method. It corresponds to the energy reaction coordinate dependence presented in figure 1. The reaction coordinate used is the projection of the CO distance into the surface plane. The threefold coordinated CO molecule starts to bend, then the oxygen end crosses a surface atom optimizing the interaction with the CO 2π* orbital and the dissociated atoms finish in positions of optimum coordination, where they share the least number of surface atoms.

The higher reactivity of the (100) surface is due to the lower coordination of the surface-atoms. This decreases delocalization of the surface atomic electron density and hence enhances the covalent interaction with adatoms (van Santen 1987), stabilizing the products. Using the same approach we also studied the association

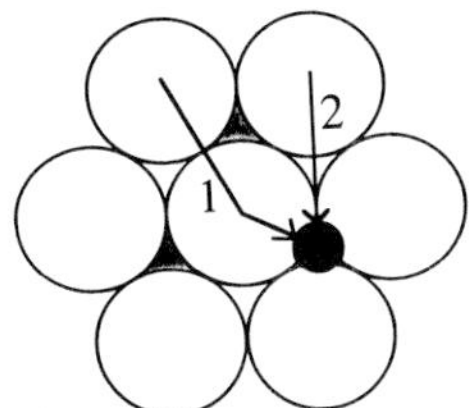

Figure 8. Recombination of CO and CH_2 with a C atom adsorbed to the Rh (111) surface (Koerts & van Santen 1991). The reaction paths show the C–C bond formation. The activation energies are, for CH_2+C, 41 kJ mol^{-1} (path 1) and 50 kJ mol^{-1} (path 2). For CO+C they are 121 kJ mol^{-1} (path 1) and 156 kJ mol^{-1} (path 2).

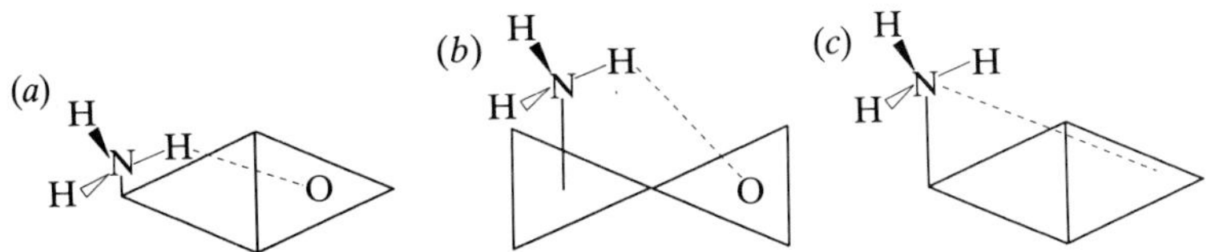

Figure 9. NH_3 oxidation on a Cu surface (Kerkhof *et al.* 1992). (*a*) No surface atom sharing reaction path between adsorbates. (*b*) One atom sharing reaction path between adsorbates. ΔH is the overall energy change of the reactions and is (*a*) 48 kJ mol^{-1}, (*b*) 48 kJ mol^{-1}, (*c*), 176 kJ mol^{-1}. E_{barr} is the maximum energy to be overcome and is (*a*) 131 kJ mol^{-1}, (*b*) 173 kJ mol^{-1}, (*c*) 344 kJ mol^{-1}.

reaction of an adsorbed carbon atom with adsorbed CH_2 and CO species, again for a Rh (111) surface (Koerts & van Santen 1991). As illustrated in figure 8, two paths were compared. In both cases the C atom starts in threefold coordination. The CH_2 fragment can approach the carbon atom by moving over one metal atom or a metal-metal bond.

In the first case one atom is shared in the transition state, in the second case two metal atoms. A significant amount of the activation energy is found to be due to the geometric necessity of the two approaching fragments to share a surface atom. The path that corresponds to sharing the least number of metal atoms is favoured.

Work by Blyholder (Blyholder & Lawless 1989) on surface-formyl formation from H and CO and Whitten (Yang & Whitten 1989, 1991), who uses first principle cluster calculations on the hydrogen addition to a chemisorbed CH_x species, gives results in line with the above conclusions. A final example concerns the dissociation of NH_3 by coadsorbed oxygen atoms on a Cu-cluster (Kerkhof *et al.* 1992). LDA calculations have been done for the different geometries denoted in figure 9. The reaction to be studied is

$$NH_3+O \rightarrow NH_2+OH.$$

The four species are adsorbed to a cluster simulating the Cu (111) surface. Dissociation is studied starting with NH_3 being adsorbed in top or threefold coordination at the same distance from the adsorbed oxygen atom. During hydrogen transfer the NH_3 and adsorbed oxygen atom never share a bond with a surface-atom when following reaction path *a*. Following reaction path *b* the NH_3 molecule and oxygen do share one surface atom. One finds that the least sharing atom reaction path has a significantly lower activation energy.

4. Conclusion

The following general rules are found to apply to chemisorption.

1. The adsorption energy decreases with increasing d-valence-electron band occupation for the group VIII transition metals.

2. The adsorption energy of a molecule is controlled by the balance of donating and backdonating terms; backdonation favours high coordination, donation can favour low coordination. The d-valence-electron interaction determines the balance.

3. Atoms usually favour high coordination. The adsorption energy decreases more strongly with d-valence-electron occupation than that of an adsorbed molecule.

For dissociation and association reactions only a few reaction paths with a low activation energy are available. For dissociative reactions, the availability of asymmetric surface-group orbitals suitable for backdonative interactions with unoccupied antibonding molecular orbitals is important. For association reactions a donative interaction which will withdraw electrons from antibonding orbitals forming between the fragments in the course of the reaction is more suitable.

A stable final configuration for the molecular fragments or atoms generated upon dissociation requires a reaction path proceeding on an ensemble of atoms, such that high coordination of the adsorbed dissociation products is possible. A large ensemble is also necessary so that reaction proceeds according to the minimum atom sharing principle.

References

Andersen, A. B. 1974 *J. chem. Phys.* **60**, 2477.

Andersen, A. B. 1975 *J. chem. Phys.* **62**, 1187.

Baerends, E. J., Ellis, D. E. & Ros, P. 1973 *Chem. Phys.* **2**, 41.

Baerends, E. J. & Ros, P. 1973 *Chem. Phys.* **2**, 52.

Baerends, E. J. & Ros, P. 1975 *Chem. Phys.* **8**, 412.

Benziger, J. B. 1980 *Appl. Surf. Sci.* **6**, 105.

Boudart, M. & Djéga-Mariadassou, G. 1984 *Kinetics of heterogeneous catalytic reactions*. Princeton University Press.

Berke, H. & Hoffmann, R. 1978 *J. Am. chem. Soc.* **100**, 7224.

Biemolt, W., Kerkhof, G. J. C. S., Davies, P. J., Jansen, A. P. J. & van Santen, R. A. 1992 *Chem. Phys. Lett.* **188**, 477.

Blomberg, M. R. A. & Siegbahn, P. E. M. 1983 *J. chem. Phys.* **78**, 5689.

Blyholder, G. & Lawless, M. 1989 *J. Am. chem. Soc.* **111**, 1275.

Brandemark, U. B., Blomberg, M. R. A., Petterson, L. G. & Sieghbahn, P. E. M. 1984 *J. phys. Chem.* **88**, 4617.

Chan, A. W. E., Hoffmann, R. & Ho. W. 1992 (In the press.)

Chattopadhyay, A., Yang, H. & Whitten, J. L. 1990 *J. phys. Chem.* **94**, 6379.

Cyrot-Lackmann, F., Desjonqueres, M. C. & Gaspard, J. P. 1974 *J. Phys.* C **7**, 925.

Ertl, G. 1983 *Catalysis 4*, ch. 3 (ed. J. R. Anderson & M. Boudart). New York: Springer.

Ertl, G. 1990 *Angew. Chem.* **102**, 1250.

Grimley, T. B. & Torrini, M. 1973 *J. Phys.* C **6**, 868.

Haydock, R., Heine, V. & Kelly, M. J. 1975 *J. Phys.* C **5**, 2845.

Hermann, K., Bagus, P. S. & Bauschlicher, C. W. 1985 *Phys. Rev.* B **31**, 6371.

Hoffmann, R. 1963 *J. chem. Phys.* **39**, 1397.

Kerkhof, G. J. C. S., Biemolt, W., Jansen, A. P. J. & van Santen, R. A. 1992 *Surf. Sci.* (In the press.)

Kiskinova, M. P. 1991 New trends in CO activation (ed. L. Guczi). *Stud. Surf. Sci. Catalysis* **64**, 37.

Kiskinova, M. P. 1992 Poisoning and promotion in catalysis based on surface science concepts and experiments. *Stud. Surf. Sci. Catalysis* **70**.

Koerts, T. & van Santen, R. A. 1991 *J. molec. Catalysis* **70**, 119.

de Koster, A. & van Santen, R. A. 1990 *Surf. Sci.* **233**, 366.

Lang, N. A. & Williams, A. R. 1978 *Phys. Rev.* B **18**, 616.

Lundquist, B. I., Gunnarson, O., Hjelmberg, H. & Norskov, J. K. 1979 *Surf. Sci.* **89**, 196.

March, N. H. 1986 *Chemical bonds outside metal surfaces*. Plenum.

Newns, D. M. 1969 *Phys. Rev.* **178**, 1123.

Norskov, J. K. 1992 *Prog. Surf. Sci.* **38**.

Norskov, J. K. & Lang, N. D. 1980 *Phys. Rev.* B **21**, 2136.

Norskov, J. K. 1982 *Phys. Rev.* B **26**, 2875.

Panas, I. & Siegbahn, P. 1988 *Chem. Phys. Lett.* **88**, 458.

Panas, I., Siegbahn, P. & Wahlgreen, U. 1989 *J. chem. Phys.* **90**, 6791.

Ponec, V. 1975 *Catalysis Rev. Sci. Engng.* **11**, 1.

Ponec, V. 1977 *Prog. Surf. Membrance Sci.* **13**, 1.

Sanchez Marcos, E., Jansen, A. P. J. & van Santen, R. A. 1990 *Chem. Phys. Lett.* **167**, 399.

van Santen, R. A. 1987 *Prog. Surf. Sci.* **25**, 253.

van Santen, R. A. 1987 *J. chem. Soc. Faraday Trans.* I **83**, 1915.

van Santen, R. A. 1991 *Theoretical heterogeneous catalysis*. Singapore: World Scientific.

van Santen, R. A. & Baerends, E. J. 1991 *Theoretical treatment of large molecules and their interactions* (ed. Z. B. Maksic), p. 323. Springer-Verlag.

van Santen, R. A. & de Koster, A. 1991 *New trends in CO activation* (ed. L. Guczi). *Stud. Surf. Sci. Catalysis*, **64**, 1.

van Santen, R. A. & Zonnevylle, M. C. 1992 *Cluster models for surface and bulk phenomena*. (ed. G. Pacchioni & P. Bagus), NATO ASI, B283. Plenum Press.

van Santen, R. A., de Koster, A. & Koerts, T. 1990 *Catalysis Lett.* **7**, 1.

Sautet, P. & Paul, J. F. 1991 *Catalysis Lett.* **9**, 245.

Schule, J., Siegbahn, P. & Wahlgren, U. 1988 *J. chem. Phys.* **89**, 6982.

Shustorovich, E. 1990 *Adv. Catalysis*, **37**, 101.

Somorjai, G. A. 1981 *Chemistry in two dimensions: surfaces*. Cornell University Press.

Thorn, D. L. & Hoffman, R. 1975 *J. Am. chem. Soc.* **100**, 2079.

Upton, T. H., Stevens, P. & Madix, R. J. 1988 *J. chem. Phys.* **88**, 3988.

Yang, H. & Whitten, J. L. 1989 *J. chem. Phys.* **91**, 126.

Yang, H. & Whitten, J. L. 1991 *Surf. Sci.* **255**, 193.

Zonnevylle, M. C., Hoffmann, R., van den Hoek, P. J. & van Santen, R. A. 1989 *Surf. Sci.* **223**, 223.

Zonnevylle, M. C., Geerlings, J. J. C. & van Santen, R. A. 1990 *Surf. Sci.* **240**, 253.

Zonnevylle, M. C. & van Santen, R. A. 1992 (In preparation.)

Discussion

P. C. H. Mitchell (*Department of Chemistry, University of Reading, U.K.*). We need to be able to validate our models of surface-bound species in catalysed reactions, for example carbon monoxide in the Fischer–Tropsch catalysis. One problem is to know whether surface structures revealed by low energy electron diffraction, for example, are participants in the catalysed reaction, or merely spectators. Here the computational approach would be to calculate the structures and reactivities of candidate surface species to guess the most probable one. We must surely start with candidates known to us from molecular chemistry. An equally important problem is to be able to validate one's computational method against species with known structures and properties (not postulated surface species). Carbonyl clusters provide a wealth of structures for computational modelling studies. Our aim in computational studies should be not only to explain observed trends but more in *delimiting possibilities*. As to whether computation is relevant, we shall only know when it has been done.

A. M. STONEHAM (*Harwell Laboratory, Didcot, U.K.*): Your cluster calculations appear to be a constant electron number. But if they represent a bulk solid, they should be at constant Fermi energy (this can make a significant difference when small energy levels are involved). But if they are for such small clusters, there won't be a Fermi surface to establish Friedell oscillations, and I wonder how representative of real clusters your one was?

R. A. VAN SANTEN. The cluster calculation for CO on Co and NH_3 on Cu have been done using the Amsterdam local density approximation code. This method gives chemically relevant predictions of bond energies as well as electron-energy distributions.

In most cases the clusters have been chosen to represent geometrically a part of a metal surface. They have not been chosen to give the minimum energy configuration for the particular cluster size studied. The Friedel oscillations refer to changes in the electron-energy distribution as a response to a disturbance. In a free-electron gas it is determined by the Fermi momentum at the Fermi surface. In a discrete solid there is also a dependence on the atom–atom distance. In the clusters there appears to be a change in the electron-energy distribution function (measured from the orbital overlap population) that has the same origin as the Friedel oscillation. If one were to adapt the Fermi level in cluster calculations to that of the bulk it implies charging of the clusters. This can be done, but makes comparison with surfaces even more difficult. Clearly one has to be very careful to extrapolate from results obtained by cluster calculations to surfaces. We have discussed this issue in an analysis of the results for the CO/Co system.

Simulation of growth processes

By J. H. Harding

Theoretical Studies Department, AEA Industrial Technology, Harwell Laboratory, Didcot, Oxon OX11 0RA, U.K.

In this paper we discuss mathematical, numerical and atomistic models of growth at surfaces through a number of examples. It is shown how such models have often been successful in predicting the structures that are observed experimentally. Different kinds of models are often complementary to each other. Mathematical models can give guidance on the long-time behaviour of the system but often need help from more detailed simulation to decide what to assume about the underlying microscopic processes. Despite their success, such models still have a long way to go. In particular, they need to address the question of how to predict the effective properties of coatings. Here experimental data are not always available, and guidance from theory is greatly needed.

1. Introduction

Interfaces are not thermodynamically stable objects. Surfaces are inevitable, but most real surfaces are not the minimum energy surfaces one could predict from a Wulff plot. Grain boundaries, voids, texture are not inevitable and a microstructure containing such features is what it is only because of the history of its making. Since the mechanisms of growth of such features depend on what is already there, this suggests that such mechanisms also depend on the detailed history of the problem. This is correct, and much effort has gone into trying to build models to explain growth at surfaces and other interfaces. Such models encounter two fundamental difficulties.

First, growth takes place on a variety of timescales. The detail of why an atom goes in one place rather than another is the proper subject of a microscopic simulation; molecular dynamics or Monte Carlo. However, the long-time behaviour may have features of its own that do not apparently depend on such details. Here a continuum approach is the more obvious starting point. Many problems are not so easily categorised; they are too slow to be tackled readily by running a detailed simulation for a long period yet the details of the process are too important to be dealt with by the crude averaging that a continuum model almost always uses. Here a wide variety of models have been used. These usually attempt to isolate some important features of the problem and incorporate them in a computer model. These are the main subject of this paper and a number of examples will be given later.

Second, the growth at a particular point on the surface may be affected by what happens far away from the growth point. A particular example of this is shadowing. If we consider a rough surface grown from a flux of particles projected onto it at an angle, parts of the 'valleys' will be shielded from the incoming particles by the

Phil. Trans. R. Soc. Lond. A (1992) **341**, 283–291
© 1992 The Royal Society
Printed in Great Britain

'mountains'. Such an effect is not difficult to treat by a computer model, but presents great difficulties for mathematical modelling because the effect is non-local. The part of the surface doing the blocking could be at a considerable distance from that being blocked.

Many of the models in the literature attempt to reproduce the observed features of the microstructure. It is clearly necessary to do this, although the fact that a model produces something that looks like experiment is not a guarantee that the model is correct. However, what is often wanted, particularly in coatings, is not the prediction of the appearance of the coating, but its properties, particularly its properties as a function of the process parameters. The properties most often wanted are the Young's modulus, Poisson ratio, thermal expansion and thermal conductivity. This requires a step beyond that taken by most modelling.

2. Solid–liquid interfaces

The growth of solids in liquids was first considered by Stefan and a vast literature has grown from it. The original Stefan problem was the advance of an ice sheet across an undercooled pond. Consider the interface between a solid and its own liquid. The properties of both phases are usually considered to be the same. The melt is undercooled and so solid forms. Heat is generated at the interface by the release of latent heat. We therefore write the equations

$$\partial T/\partial t = \Lambda \nabla^2 T \tag{2.1}$$

and, at the liquid–solid interface, we have

$$[-C_p \Lambda \hat{n} \nabla T] = v_{\hat{n}} L, \tag{2.2}$$

where Λ is the thermal diffusion coefficient, C_p the specific heat, L the latent heat of fusion and $v_{\hat{n}}$ the velocity of advance of the interface where $\hat{n}$ is the normal to the interface pointing into the liquid. In general the interface will be curved. To discuss this, we must add an assumption about the microscopic processes at the interface. The normal assumption is that they are so fast that the system can be assumed to be in local thermodynamic equilibrium. This gives a relation between the temperature at a curved interface, T_c, and the temperature at the planar interface, T_p:

$$T_c = T_p(1-\gamma\kappa/L), \tag{2.3}$$

where γ is the surface tension and κ is the inverse curvature. What then is the rate of advance of the interface? The solutions to this problem are discussed by Kessler *et al.* (1988) and Brener & Mel'nikov (1991). We do not consider here the detailed mathematical treatments of the problem (for which see the references above) but make a few qualitative comments on the results.

The planar case is completely unstable. This is because, if bumps spontaneously form, they will lose heat more effectively than the surroundings and therefore grow. Ivantsov presented an approximate solution; that of a growing parabolic tip. Dendrites do have approximately this shape, but this solution is still unstable. Further progress has only been possible by simplifying the problem using a local approximation (discussed by Kessler *et al.*). This shows the fundamental importance of surface tension which, in effect, selects only the fastest growing of the solutions and can explain why dendritic structures occur. The Ivantsov problem is of such interest because it shows the wide range of behaviour that can result from so simple a system. The concentration of mathematical effort has led to an understanding of the structure of the growths, provided simplifying assumptions are made.

We have considered the Ivantsov equations as a continuum model. They are also the basis of a numerical model; diffusion-limited aggregation (see Meakin (1987) for an introduction and original references; most work is based on Witten & Sander (1981)). This model considers a cluster in a medium of particles executing random walks. If they hit the cluster they stick to it. The diffusion equation for the probability of an incoming particle is therefore

$$\partial u/\partial t = D\nabla^2 u, \tag{2.4}$$

where D is the diffusion coefficient. The boundary conditions are $u(r,t) = 0$ at the cluster surface and $u(r_\infty, t) = u_\infty$ far away from the cluster. If the cluster is a solid of density ρ, the average velocity of advance normal to the surface, v_n is given by $\rho v_n = D\,\partial u/\partial n$. These are the Ivantsov equations without surface tension. The instabilities noted there are suppressed by two cut-offs. The finite size of the lattice on which the simulation is performed gives a lower cut-off; the diffusion length gives an upper one. Noise is introduced into the system by the fact that individual particles are deposited. This is amplified into the characteristic 'fingers' seen in such growths by the instabilities. Indeed, Honjo *et al.* (1986) have produced a system that passes from the dendritic form to the 'fingers' of the diffusion-limited aggregation model. Matsushita *et al.* (1984, 1985) have shown that the patterns produced in the electro-deposition of zinc strongly resemble the simulations made using this model. Unfortunately, the only point of comparison given is the fractal dimension.

3. Solid–gas interfaces

These interfaces have frequently been studied by atomistic simulation. Direct simulation of growth processes has also been attempted; molecular dynamics simulations of molecular beam epitaxy have been reviewed by Das Sharma (1988). We shall not consider these in detail, but continue to discuss the approach used in §2. There are, however, two important differences from those cases. The first is that the initial surface is given; it is the substrate on which the coating is made. The second is that coatings are made by a far wider variety of methods than those seen in liquid/surface interfaces. As one might expect, the microstructure of the result depends on the method used to form it. Even using one method, the microstructure can vary greatly depending on the precise conditions used. We shall consider two examples, sputtering and molecular beam epitaxy.

Thornton (1977) divides the microstructure of coatings produced by sputtering into four regions characterized by the ratio of the temperature of the substrate to the bulk melting temperature of the coating; T/T_{M}. For $T/T_{\mathrm{M}} < 0.3$, the microstructure is columnar with fine grains but also many voids. At higher temperatures, $0.3 < T/T_{\mathrm{M}} < 0.5$, surface diffusion becomes important and the coating has larger grains and fewer gross defects. Beyond this region, bulk diffusion and recrystallization occur producing a structure resembling that of material grown from the melt. A transition zone is often observed in the region of $T/T_{\mathrm{M}} = 0.3$.

The most obvious feature of the microstructure of these coatings is the columns. These often obey the *tangent rule*. If α is the angle of incoming particles to the normal to the interface and β the angle of the columns in the coating, $2\tan\beta = \tan\alpha$. Columns are only observed for sizeable values of α. For near-normal incidence a 'cauliflower' structure is obtained (Messier & Yehoda 1985). For very high values of the angle of incidence, the tangent rule fails (Bensiman *et al.* 1984); a better approximation is $\beta = \alpha - 17$ (degrees).

Sputtered coatings are made by projecting particles at the surface. It is not difficult to model this. The ballistic model throws particles at a surface; when they hit the surface they stick to the place they strike. Henderson *et al.* (1974) showed that this model produces a set of columns and voids that resemble sputtered coatings. It even predicts the tangent rule (Leamy & Dirks 1978). Unfortunately, as Kim *et al.* (1978) pointed out, the scale of the predicted microstructure is wrong by several orders of magnitude. It is a good illustration of a general point made by Meakin (1987). It is necessary to produce a microstructure that looks right; it certainly is not sufficient. In this case, the problem arises because the model assumes that when an incoming particle hits the growing coating, it sticks where it hits. In fact, measurements of the sticking coefficient (S. J. Bull, personal communication) give values of about 0.1 under the conditions where coatings are formed. Simulations have been performed with low values of the sticking coefficient (Kim *et al.*). These do give more reasonable values for the density; however, as the sticking coefficient is lowered, the columns become less visible although a more detailed analysis of the results shows that they are still there. The fractal dimension of the coating is unaltered (Meakin 1983).

These results show that the behaviour of the impinging particles is not the only factor in the growth of coatings. Completely different behaviour has been observed when surface diffusion is dominant. Van der Drift (1967) has argued that this, together with the assumption that only the growth rate of crystallites normal to the interface is important, can explain the morphology of a wide range of coatings. Crystallites with the fastest growing orientation perpendicular to the interface then dominate the structure (platelets 'stand on end' for example). The fastest growing orientation is not the same as the fastest growing surface; it is well known that such surfaces 'grow out'.

This suggests that a growth model must contain both the manner in which particles arrive at the surface and a surface diffusion term (or at least some term that anneals out the instabilities produced by the simple growth model). This, however, says nothing specific about the geometry of the surface. If we are building a numerical model, this will be dealt with automatically. However, if we are using a mathematical model, something specific must be done. In the case of sputtered films, the most obvious geometrical effect is shadowing. The vapour atoms come in from all directions, and the local growth rate is roughly proportional to the solid angle of exposure of the site. The general shadowing effect is very difficult to include in a mathematical model, but more local effects of surface shape can be introduced. An example of this is the explanation of the tangent rule by Leamy & Dirks (1978).

The Eden model (see Bruinsma *et al.* (1990) for references) assumes that the local surface grows in the direction normal to itself at a given rate. If we assume that the coating is grown on a flat substrate, this implies that growth perpendicular to the substrate is faster at slopes in the surface than at hills or valleys. This is the basis of the KPZ equation (Karder, Parisi & Zhang 1986). This is a growth law for the height of the coating, h,

$$\partial h/\partial t = \sigma \nabla^2 h + J + \lambda(\nabla h)^2 + \eta(r, t), \tag{3.1}$$

where σ is the surface energy, λ a coefficient, J the rate of arrival of material and $\eta(r,t)$ represents the noise caused by the arrival of particles (as in the diffusion-limited aggregation model). The first term has an annealing effect; this could be mediated either by evaporation/condensation or reconstruction (Edwards & Wilkinson 1982). The third term is the lowest order contribution of the Eden model. The equation also incorporates the solid-on-solid approximation which states that the system contains no overhangs (see Conrad (1992) for an extended discussion of the approximation and its use in simulations).

This model, and developments of it, has been studied in great detail. Since it does not contain any effects of shadowing, it is not suitable for sputtering and cannot predict columnar growth. Although the model would seem closer to molecular beam epitaxy, Das Sharma & Tamborenea (1991) have shown that this is only superficially so. The crucial difference is that any relaxation effects in such models are instantaneous, whereas models of molecular beam epitaxy have a further timescale imposed on them by the effects of surface diffusion. In a later paper, Lai & Das Sharma (1991) suggest that a possible model equation for molecular beam epitaxy is

$$\partial h/\partial t = -\lambda_1 \nabla^4 h + \lambda_2 \nabla^2 (\nabla h)^2 + \eta(r,t). \tag{3.2}$$

Both the first two terms can be interpreted as modelling competition between surface diffusion and bonding effects. An equation like this shows both the strength and the weakness of the mathematical approach. It is possible to analyse an equation like this to obtain general features of the growth; in this case the scaling relations among the growth exponents. However, there is no unique connection between the form of the terms in the equation and any physical process and no way of predicting the relative importance of various terms.

As an example of how one can produce similar equations by quite different mechanisms, let us turn back to the question of shadowing in sputtered films. Karunasiri *et al.* (1989) have considered the following model equation

$$\partial h/\partial t = -D\,\partial^4 h/\partial r^4 + J\theta(r,h) + \eta(r,t). \tag{3.3}$$

The first term represents the surface diffusion; D is the surface diffusion coefficient. The second term is the incoming particle flux where $\theta(r,h)$ is the angle of exposure of a site at r,h and $\eta(r,t)$ is the noise term. The solid-on-solid approximation is used here as above. If the shadowing term dominates, the system evolves into an approximately self-similar 'mountain' landscape. Whether this happens depends on the ratio D/J. For large enough values, it should be possible to grow flat coatings. As the authors note, the restriction to one dimension may exaggerate the shadow effect, but the idea of a transition to a rough surface is not implausible. Mazor *et al.* (1988) came to similar conclusions by their Monte Carlo model. They also investigated the coarsening of the microstructure of the film using Monte Carlo methods to be discussed in the next section. The results illustrate the argument of Van der Drift that the surfaces growing fastest in the vertical direction dominate the problem.

4. Solid–solid interfaces

Growth is not just something that occurs at surfaces. In this section we consider the simulation of grain boundary growth. Here, as before, we are not attempting to calculate the properties of ions at individual grain boundaries, but rather are using models to gain understanding of the behaviour of the system as a whole. The experimental position has been reviewed by Atkinson (1988). The basic result is that for isothermal annealing, the mean grain size $\bar{R}$ grows as

$$\bar{R} = kt^n \tag{4.1}$$

at long times. The constant k has an Arrhenius dependence on temperature, but the activation energy is not identifiable with any simple atomic process. n is the grain boundary exponent. It is the purpose of most theories to calculate this.

The grain boundary network coarsens because it is always energetically favourable to eliminate grain boundaries. The problem is, however, complicated by the topology of the network; the way a given boundary moves is affected by the behaviour of other boundaries connected to it. Theories to explain the behaviour are of two kinds; the mean field theories, such as those of Hillert (1964) and Louat (1974) and the Monte Carlo simulations developed by Anderson *et al.* (1984). We will approach the problem by way of the simulations.

In this method the microstructure is modelled on a mesh. Each mesh point is assigned a spin; these spins will label the different possible orientations of a grain in a growing lattice. There must be a large enough number of possible spin orientations so that grains of like orientation do not meet each other. In practice 30–40 orientations is enough. The grain boundary energy is modelled using the many-spin Potts hamiltonian

$$H = -J \sum_{NN} (\delta_{S_i S_j} - 1) \tag{4.2}$$

where J is a parameter (in principle temperature dependent) and the sum is over all nearest neighbours. The system is then allowed to evolve by flipping the spins using standard Monte Carlo techniques. It is important to ensure that the system has reached the true long-time behaviour. Early work did not do this and obtained growth exponents of about 0.4, in disagreement with models like those of Hillert (1964) which assumed that the driving force for coarsening was the reduction of grain boundary energy. Later simulations (Anderson *et al.* 1989) obtained exponents of 0.5 in agreement with Hillert.

Hillert's theory solves the continuity equation for the probability distribution in a network, given the equation of motion of the radii of the grains. This he took to be

$$v(r,t) = \partial r/\partial t = c(u)/\bar{r} - 1/r \tag{4.3}$$

where $u = r/\bar{r}$. The first term represents grain growth due to the influx of material from neighbouring grains; the second term represents grain shrinkage by surface tension. Hillert took $c(u) = 1$, a mean field theory for all grains. However, the simulations of Frost *et al.* have shown deviations from mean field behaviour; large grains cluster near small grains and vice versa. Mulheran (1992*a*) has analysed the results from simulations to suggest that a reasonable approximation to $c(u)$ is simply $0.6u$. This gives much better agreement with the frequency distribution functions obtained from simulations.

Mulheran (1992*b*) has argued that a different mechanism drives grain growth at short time scales and has investigated the random walk theory of Louat (1974). This suggested that the evolution of the network was by the random motion of the grain boundaries. Mulheran showed that Louat's conclusion that the growth exponent was 0.5 is true only in the one-dimensional case; in two dimensions the exponent is $\frac{1}{3}$ and in three dimensions $\frac{1}{4}$. Since simulations obtain low growth exponents for short runs, this suggests that both mechanisms might operate in this region. By using the grain migration model of Feltham (1957) Mulheran has shown that this is a reasonable conclusion.

Simulations of this kind have been used to probe a wide variety of microstructural behaviour in solids. Their main use to date has been to test the assumptions behind the mean field theories often used, and suggest improvements. The most obvious difficulty is that, since there is no reliable method of fixing the value of J, their predictive power is limited.

5. Obtaining effective properties from simulations

It is known that the microstructure affects the properties of materials, and there are many comments to this effect (see, for example, Dirks & Leamy 1977). However, detailed studies of the variation of properties with microstructural features are rare. Examples are given by Hoffman *et al.* (1991) and Håkansson *et al.* (1991). In this section we consider briefly how knowledge of the microstructure can be used to obtain material properties. The theories that have been evolved face an awkward choice; either one performs a very crude average (in which case the effect of the microstructure is washed out) or one attempts to perform a complete analysis (which requires so much data about the details of the microstructure that it is almost impossible to obtain it). The cases of thermal conductivity and elastic properties illustrate the dilemma.

Maxwell (1873) showed that for a dilute two-phase system containing spherical inclusions, the effective thermal conductivity, K, is given by

$$K/K_0 = (1-p)/(1+0.25p), \tag{5.1}$$

where K_0 is the thermal conductivity of the pure material and p is the volume fraction of the inclusions. This result ignores both the shapes of the inclusions and their possible interaction. More recent work has attempted to consider inclusions of arbitrary shape. Here most of the interest has been in the effect of porosity. A collection of the most commonly used results is given by Schulz (1981). As she notes, for highly porous materials, it is not enough to know gross features such as the pore fraction; the detailed structure of the material becomes important. Such effects also appear in empirical attempts to correlate porosity and thermal conductivity (Ebel & Vollath 1988). A similar point is made by Lu & Kim (1990) in their analysis of the effective thermal conductivity of composites where the inclusions have very high conductivity. To obtain their results to order p^2, they require the pair distribution function for the inclusions. This is only obtainable analytically for simple model systems. It is not easy to obtain from experiment and has rarely, if ever, been attempted.

A similar line of argument obtains for elastic properties and stresses. A review of the effect of microstructure on stresses on thin films is given by Doerner & Nix (1988). The first level of approximation is again to consider only the volume fractions of the constituents. A review of the various results obtainable is given by Hale (1975). Beyond that, information about the inclusion distribution functions is required. If this is available, a method of obtaining effective elastic properties is discussed by Ferrari & Johnson (1989). However, now the full distribution function of the inclusions is needed. This method has been used, together with a numerical model, to obtain effective elastic properties and stresses for a plasma-sprayed zirconia coating (Ferrari *et al.* 1991). This suggests that the use of a model can overcome the major difficulty with using sophisticated effective medium approaches; it can provide the data about the microstructure that such theories require.

6. Conclusions

Many different approaches have been used in the attempt to simulate the growth at surfaces. The various methods used are often complementary. Mathematical models are often used to check the long-time behaviour of simulations. The simulations can give guidance on what mathematical models should assume about the underlying behaviour. In some cases the numerical models can pose the kinds of questions that can be tackled by atomistic simulation; the question of sticking coefficients or the mechanism of grain boundary motion for example.

Most models attempt to predict the structure of interfaces or coatings. However, as we have tried to suggest, the prediction of patterns and structure, though useful, is not sufficient. If these models are really to be useful to those attempting to produce coatings, they need to predict properties. There are a number of cases where this is now being done. More are needed, both to test the models against hard data and to help experimentalists to produce better coatings.

References

Anderson, M. P., Srolovitz, D. J., Grest, G. S. & Sahni, P. S. 1984 Computer simulation of grain growth. I. Kinetics. *Acta metall.* **32**, 783–791.

Anderson, M. P., Grest, G. S. & Srolovitz, D. J. 1989 Computer simulation of normal grain growth in three dimensions. *Phil. Mag.* B **59**, 293–329.

Atkinson, H. V. 1988 Theories of normal grain growth in pure single phase systems. *Acta metall.* **36**, 469–491.

Bensiman, D., Shraiman, B. & Liang, S. 1984 On the ballistic model of aggregation. *Phys. Lett.* A **102**, 238–240.

Brener, E. A. & Mel'nikov, V. I. 1991 Pattern selection in two-dimensional dendritic growth. *Adv. Phys.* **40**, 53–97.

Bruinsma, R., Karunasiri, R. P. U. & Rudnick, J. 1990 Growth and erosion of thin solid films. In *Kinetics of growth and ordering at surfaces* (ed. M. G. Lagally), pp. 395–413. New York: Plenum.

Conrad, E. H. 1992 Surface roughening, melting and faceting. *Prog. Surf. Sci.* **39**, 65–116.

Das Sharma, S. 1989 Numerical studies of epitaxial kinetics: what can computer simulation tell us about nonequilibrium crystal growth? *J. Vacuum Sci. Tech.* A **8**, 2714–2726.

Das Sharma, S. & Tamborenea, P. 1991 A new universality class for kinetic growth: one dimensional molecular-beam epitaxy. *Phys. Rev. Lett.* **66**, 325–328.

Dirks, A. & Leamy, H. J. 1977 Columnar microstructure in vapor-deposited thin films. *Thin Solid Films* **47**, 219–233.

Doerner, M. F. & Nix, W. D. 1988 Stresses and deformation processes in thin films on substrates. *CRC Crit. Rev. Solid State Mater. Sci.* **14**, 225–268.

Ebel, H. & Vollath, D. 1988 Experimental correlations between pore structure and thermal conductivity. *J. nucl. Mater.* **153**, 50–58.

Edwards, S. F. & Wilkinson, D. R. 1982 The surface statistics of a granular aggregate. *Proc. R. Soc. Lond.* A **381**, 17–31.

Feltham, P. 1957 Grain growth in metals. *Acta metall.* **5**, 97–105.

Ferrari, M., Harding, J. H. & Marchese, M. 1991 Computer simulation of plasma-sprayed coatings. II. Effective bulk properties and thermal stress calculations. *Surf. Coat. Tech.* **48**, 147–154.

Ferrari, M. & Johnson, G. C. 1989 Effective elasticities of short-fiber composites with arbitrary orientation distribution. *Mech. Mater.* **8**, 67–73.

...ITY SCIENCE LIBRARY

Frost, H. J., Thompson, C. V. & Walton, D. T. 1990 Simulation of thin film grain structures. I. Grain growth stagnation. *Acta metall.* **38**, 1455–1465.

Håkansson, G., Hultman, L., Sundgren, J.-E., Greene, J. E. & Münz, W.-D. 1991 Microstructures of TiN films grown by various physical vapour deposition techniques. *Surf. Coat. Tech.* **48**, 51–67.

Hale, D. K. 1976 The physical properties of composite materials. *J. mater. Sci.* **11**, 2105–2141.

Henderson, D., Brodsky, M. H. & Chaudari, P. 1974 Simulation of structural anisotropy and void formation in amorphous thin films. *Appl. Phys. Lett.* **25**, 641–643.

Hillert, M. 1964 On the theory of normal and abnormal grain growth. *Acta metall.* **13**, 227–238.

Hoffman, R. A., Lin, J. C. & Chambers, J. P. 1991 The effect of ion bombardment on the microstructure and properties of molybdenum films. *Thin Solid Films* **206**, 230–235.

Honjo, H., Ohta, S. & Matsushita, M. 1986 Irregular fractal-like crystal growth of ammonium chloride. *J. phys. Soc. Japan* **55**, 2487–2490.

Karder, M., Parisi, G. & Shang, Y.-C. 1986 *Phys. Rev. Lett.* **56**, 889–891.

Karunasiri, R. P. U., Bruinsma, R. & Rudnick, J. 1989 Thin-film growth and the shadow instability. *Phys. Rev. Lett.* **62**, 788–791.

Kessler, D. A., Koplik, J. & Levine, H. 1987 Pattern formation far from equilibrium: the free space dendrite crystal. In *Patterns, defects and microstructures* (ed. D. Walgraef), pp. 1–11. Dordrecht: Nijhoff.

Kessler, D. A., Koplik, J. & Levine, H. 1988 Pattern selection in fingered growth phenomena. *Adv. Phys.* **37**, 255–339.

Kim, S., Henderson, J. & Chaudari, P. 1977 Computer simulation of amorphous thin films of hard spheres. *Thin Solid Films* **47**, 155–158.

Lai, Z.-W. & Das Sharma, S. 1991 Kinetic growth with surface relaxation: continuum versus atomistic models. *Phys. Rev. Lett.* **66**, 2348–2351.

Leamy, H. J. & Dirks, A. G. 1978 Microstructure and magnetism in amorphous rare-earth transition-metal thin films. I. Microstructure. *J. appl. Phys.* **49**, 3430–3438.

Louat, N. P. 1974 On the theory of normal grain growth. *Acta metall.* **22**, 721–724.

Lu, S.-Y. & Kim, S. 1990 Effective thermal conductivity of composites containing spheroidal inclusions. *A.I.Ch.E. J.* **36**, 927–938.

Matsushita, M., Sano, M., Hayakawa, Y., Honjo, H. & Sawada, Y. 1984 Fractal structures of zinc metal leaves grown by electro-deposition. *Phys. Rev. Lett.* **53**, 286–289.

Matsushita, M., Hayakawa, Y. & Sawada, Y. 1985 Fractal structure and cluster statistics of zinc metal trees deposited on a line electrode. *Phys. Rep.* A **32**, 3814–3816.

Maxwell, J. C 1873 *A treatise on electricity and magnetism*, vol. 1. Oxford: Clarendon Press.

Mazor, A., Srolovitz, D. J., Hagan, P. S. & Bukiet, B. G. 1988 Columnar growth in thin films. *Phys. Rev. Lett.* **60**, 424–427.

Meakin, P. 1983 Diffusion-controlled cluster formation in two, three and four dimensions. *Phys. Rev.* A **27**, 604–607.

Meakin, P. 1987 Fractal scaling in thin film condensation and material surfaces. *CRC Crit. Rev. Solid State Mater. Sci.* **13**, 143–187.

Meakin, P. 1990 Computer simulation of growth and aggregation processes. In *Kinetics of growth and ordering at surfaces* (ed. M. G. Lagally), pp. 111–135. New York: Plenum.

Messier, R. & Yehoda, J. E. 1985 Geometry of thin film morphology. *J. appl. Phys.* **58**, 3739–3745.

Mulheran, P. A. 1992*a* Mean field simulations of normal grain growth. *Acta metall.* **40**, 1827–1833.

Mulheran, P. A. 1992*b* The scaling theory of normal grain growth: beyond the mean-field approximation. *Phil. Mag. Lett.* **65**, 141–145.

Schulz, B. 1981 Thermal conductivity of porous and highly porous materials. *High Temp. High Press.* **13**, 649–660.

Thornton, J. A. 1977 High rate thick film growth. *A. Rev. mater. Sci.* **7**, 239–260.

Van der Drift, A. 1967 Evolutionary selection: a principle governing growth orientation in vapour-deposited layers. *Philips Res. Rep.* **22**, 267–278.

Witten, T. A. & Sander, L. M. 1981 Diffusion-limited aggregation; a kinetic critical phenomenon. *Phys. Rev. Lett.* **47**, 1400–1403.

Electronic structure of surfaces and of adsorbed species

By J. B. Pendry

The Blackett Laboratory, Imperial College of Science, Technology and Medicine, London SW7 2BZ, U.K.

Electrons are the main probe for determining surface crystallography. Existing methods have already established an impresive list of completed structures but, such is the demand for structural information at surfaces, new ways of interpreting diffraction data are being explored with a view to extending the power and flexibility of tools available to us.

1. Electronic structure and crystallography

Electronic structure has many roles to play at the surface: bonding, transport, spectroscopy, and determination of surface crystallography. In this paper I shall concentrate on the last of these. The arrangement of atoms at surfaces must be their most fundamental property, and until we can provide that we have no microscopic description of the surface. Although it is true that bonding provided by valence electrons determines the crystallography, as a means of actually determining the arrangement of atoms, electronic structure calculations remain an ideal just as they are in the bulk of a solid. That is not to say that they do not have a fundamental role, simply that their role is one of explanation and understanding rather than of providing details of individual structures. Elucidation of surface crystallography is largely done by diffraction of electrons with energies in the 50 eV to 500 eV range, and by subsequent interpretation of the often complex data.

Determination of surface crystallography is so important an issue that many different approaches have been tried. One of the earliest to give atomic positions was field-ion microscopy (Müller & Tsong 1969), which images atoms by ionizing helium in the intense electric fields generated by a sharp tip held at a high voltage. Remarkable images of surfaces are seen, but the technique languishes under several disadvantages: only extremely strong materials can withstand the strong electric fields, and even in these cases there is debate about whether the fields distort atomic arrangements. Only the top layer of atoms is visible and we now know that surface rearrangements can take place several layers deep into the surface. These disadvantages combined with the low resolution of the technique prevent it from occupying centre stage, but nevertheless elegant contributions are made by this technique from time to time (Fink 1986).

Sharp tips made a reappearance in surface crystallography with the scanning tunnelling microscope (STM) (Binnig & Rohrer 1982). This technique has taken the world of surface science by storm: unlike the field ion microscope (FIM) STM can study a wide range of materials, and it has the great advantage of producing real-space images of the surfaces which can be viewed and interpreted almost instantaneously. Overwhelmingly its application has been to surface *topography* rather than to

Phil. Trans. R. Soc. Lond. A (1992) **341**, 293–300
© 1992 The Royal Society

crystallography, making a vital contribution on this mesoscopic scale where diffraction methods run into problems of interpretation. As a surface crystallographic technique it has many of the disadvantages of the FIM: restriction to the outer layer of the surface and low resolution. The STM image of a single atom adsorbed on a surface is at least 5 Å† across, and interpretation of structure on a finer scale than this involves some wishful thinking. Yet in many ways the STM is an ideal companion to diffraction techniques, filling the gaps they leave.

The workhorses of surface crystallography are low-energy electron diffraction, photoelectron diffraction, high-energy ion scattering, and X-ray diffraction. Of the known surface structures, the overwhelming majority have been determined by low-energy electron diffraction.

2. LEED: the workhorse of surface crystallography

Electrons with energy in the range 50–500 eV penetrate only 5–10 Å into the surface, presenting us with ideal sensitivity for surface crystallography: most surfaces become bulk-like at depths greater than these. Their wavelength is around 1 Å, enabling resolutions approaching 0.01 Å to be achieved in deal circumstances. Traditional methodology involves acquisition of data for several diffracted beams over a range of several hundred eV, with the assumption that the surface has a well-ordered periodic structure (Pendry 1974; Van Hove & Tong 1979; Heinz & Müller 1982). The complex diffraction intensities contain all the information necessary to find the structure: it is doubtful whether any other technique is capable of generating such a volume of information. Unfortunately extraction of this information is not trivial and proceeds via trial and error. We guess a structure, calculate the diffracted intensities and compare with experiment. The latter step is done by means of an R-factor measuring overall quality of agreement with experiment for the trial structure. Agreement is never perfect, but R-factors as low as 0.25 represent 'good agreement' with experiment, on a scale where 0 is perfect agreement and 1 is no agreement at all. This method is effective for simple surfaces and something like 500 to 1000 structures have been determined in this fashion (MacLaren *et al.* 1987). With increasing complexity trial and error becomes more and more time-consuming, as the possibilities grow exponential with system size. Nevertheless it has proved possible to determine some fairly complex structures.

An early example by Andersson & Pendry (1980) is shown in figure 1: CO is adsorbed on a Cu(100) in a $C(2\times 2)$ structure. This was the first surface molecular structure to be determined, and is typical of the molecular phase of interaction with surfaces. The CO molecule is relatively weakly bound, perched high above the surface on top of a copper atom as befits a molecule whose bonding arrangements are already well satisfied. The molecular bond length at the surface is within experimental error of the gas-phase value, also reflecting the weak nature of the bonding. In contrast, atomic carbon or atomic oxygen typically bury themselves deeply in the surface, presumably to saturate their bonding requirements.

Similar sorts of results can be achieved with photoelectron diffraction and surface EXAFS, both of which share the electron as active ingredient, and have similar advantages and disadvantages.

Surface crystallographers are presently in the position that protein crystallographers were some years ago: their experimental data contain all the

† 1 Å = 10^{-10} m = 10^{-1} nm.

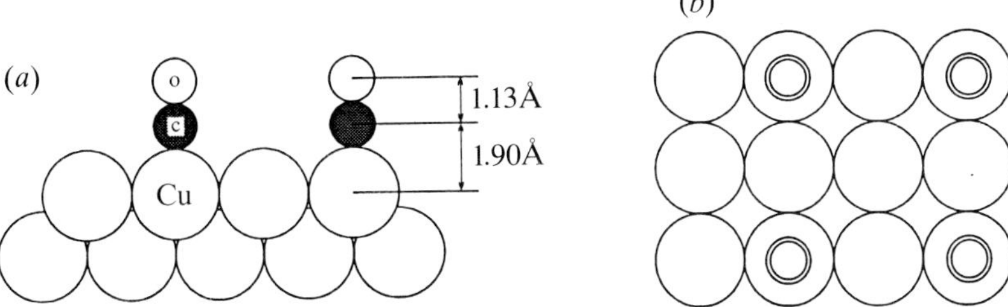

Figure 1. The structure of C(2 × 2) CO adsorbed on a Cu (100) surface after Andersson & Pendry (1980): (*a*) side view and (*b*) top view. The molecule adsorbs carbon end down, standing vertically on top of a copper atom.

information necessary to determine the surface structure, but inadequacies in methodology stand in the way of comprehensive exploitation of this information. Much development in the field in the past few years has been directed to removing this restriction, with some notable successes, and many ideas for future improvements.

3. Multiple scattering: the dragon of surface crystallography

If surface crystallography is imprisoned in the data by the dragon of multiple scattering, who will play St George?

One candidate is the *tensor-LEED* method developed by Philip Rous and myself (Rous 1992). Its original aim was to speed comparison of a given trial structure to experiment, but it has also spawned other more radical approaches to the problem. At the core of tensor-LEED lies the following idea. Suppose we make a reasonable guess at the trial structure, calculate the expected diffraction intensities, and compare with experiment, but find that the fit is not perfect. In the old method we simply start again with a new trial structure, but in tensor-LEED the first trial calculation is used to make a perturbation expansion with the original structure as a *reference structure*. Much of the computational work has already been done, and the perturbation can be accomplished very rapidly. The vicinity of a given structure can be explored at little computational cost. For structures where many coordinates are to be adjusted, some perhaps by only a small amount, this can be extremely valuable. An instance might occur when an adsorbate reconstructs the substrate. Motion of substrate atoms is often vital in obtaining an accurate and reliable structure, yet often substrate atoms move by no more than 0.1 Å. An example is given in figure 2, showing the P(2 × 2) C_2H_3 structure on an Rh (111) surface (Wander *et al.* 1991, 1992). In each unit cell there are two adsorbate atoms (discounting the hydrogen) and eight substrate atoms whose positions are to be adjusted. Without tensor-LEED this would be an almost impossible task, but it can now be accomplished in a routine manner.

One further step has been taken in developing tensor-LEED. Perturbation theory, on which the whole idea rests, expresses the change in diffracted amplitudes in terms of a matrix element of the change in scattering potential,

$$\delta A_g = \sum_j Y\langle \boldsymbol{k}+\boldsymbol{g}; \mathrm{out} \,|\, \delta t(\delta \boldsymbol{R}_j) \,|\, \boldsymbol{k}; \mathrm{in}\rangle = \sum_j M_{gj} f(\delta \boldsymbol{R}_j), \tag{1}$$

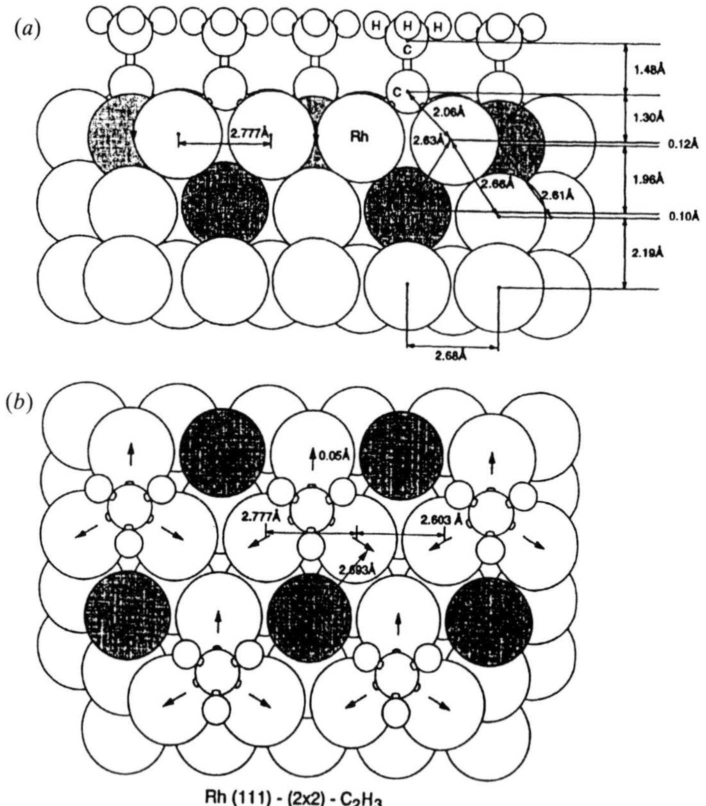

Figure 2. The structure of $P(2 \times 2)\,C_2H_3$ adsorbed on an Rh (111) surface after Wander *et al.* (1991): side view (top panel) and top view (lower panel). Note the reconstruction in the Rh substrate.

where $f(\delta \boldsymbol{R}_j)$ is a function of the displacement of the jth atom and M_{gj} is a matrix, or tensor, that is independent of $\delta \boldsymbol{R}_j$. Once M_{gj} is computed for a reference structure, $f(\delta \boldsymbol{R}_j)$ can easily be calculated for each new structure and diffracted amplitudes evaluated by simple matrix multiplication. Hence the original power of the method. Aside from simplicity, equation (1) has another property: there is a linear relation between the structural information contained in the $f(\delta \boldsymbol{R}_j)$ and the diffracted amplitudes, δA_g. In essence we can say that each atom has a characteristic contribution to δA_g which is linearly independent of the other atomic displacements.

This concept of linear independence opens new methods for interpretation of data. To see why it is so important consider a simple analogy. Suppose that we lose the keys to a safe containing some vital information. One possibility would be to call the manufacturers and borrow their bunch of replica keys. This could prove a very time-consuming solution: suppose that the lock has three tumblers and each has ten different positions, there would be 10^3 trials to be made. The other possibility is to contact a locksmith with a set of skeleton keys. He will need three, one for each tumbler, and provided that each tumbler is independent of the rest he will have ten possibilities to try for each tumbler, a total of only 30 for the lock as a whole. He will open the safe before we have barely started on the bunch of replica keys!

The moral for our problem is that we must extract independent contributions to the diffracted amplitudes and fit them one by one. In some experiments this is easily

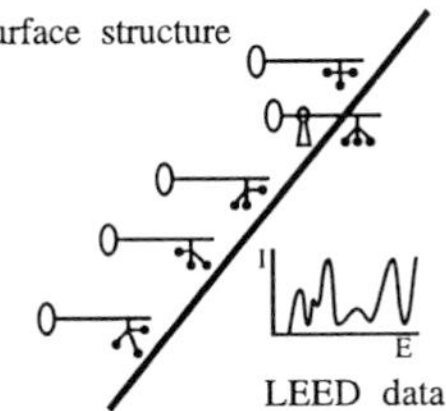

Figure 3. Traditionally LEED data are interpreted by trial-and-error methods which become time-consuming for complex structures. Adjusting the coordinates of atoms in a trial-and-error process is time-consuming, particularly when there are many atoms.

done. For example, in an EXAFS experiment atoms in each shell contribute a given Fourier component of the absorption cross-section, which can be extracted by Fourier transformation and its intensity related to occupancy of that shell. We observe that sine waves are not the only waveform that can be extracted from data, and even more complex signals occurring in multiple scattering can be extracted. The overriding requirement is one of linear independence.

Some progress towards this objective has been made in developments of direct methods (Pendry & Heinz 1990) and exploited by Wander *et al.* (1992) to simplify structure determination. In a recent paper they took the $P(2\times 2)\,C_2H_3$ structure on a Pt (111) surface, similar to the corresponding Rh (111) structure shown in figure 2, as a test bed for their calculations. The diffracted intensities were calculated for a zero-order reference structure that consisted of the unreconstructed substrate with the $P(2\times 2)\,C_2H_3$ adsorbed on top. In addition three other hypothetical structures were considered: in the first the $P(2\times 2)\,C_2H_3$ was raised by 0.5 Å, in the second the uncoordinated surface Pt was moved by 0.5 Å, and in the third both the $P(2\times 2)\,C_2H_3$ and the Pt were moved by 0.5 Å. To test whether the principle of linearity holds for this system, the intensities for the third structure were also calculated by adding together the changes in amplitudes found for structures one and two, on the assumption that the amplitudes combine linearly. The resulting diffracted intensities were compared with the conventional calculation which makes no assumptions and found to agree extremely well: the R-factor was less than 0.09. This new method, linear LEED as it has been christened, adds yet another refinement to the speed of conventional analysis while offering the possibility of more direct analysis of the data.

4. Future possibilities

The techniques that we have discussed so far are part of the present reality of structure analysis, in the process of being applied to numerous complex structure determinations. More exotic possibilities are under investigation which if realized would transform surface crystallography, but have some way to go before they are accepted as a practical means of data analysis.

Szöke (1986) followed by Barton (1988) first proposed an analogy between photoemitted electrons and holography. Figure 4 shows the principles involved in holography. Compare figure 4 with figure 5, where we show a photoemission experiment in which an electron is ejected from the inner core of an adsorbed atom. The outgoing wave divides itself between a component heading directly out of the surface, and a second component that first scatters from atoms in the later below. Interference between the two components generates a hologram-like structure.

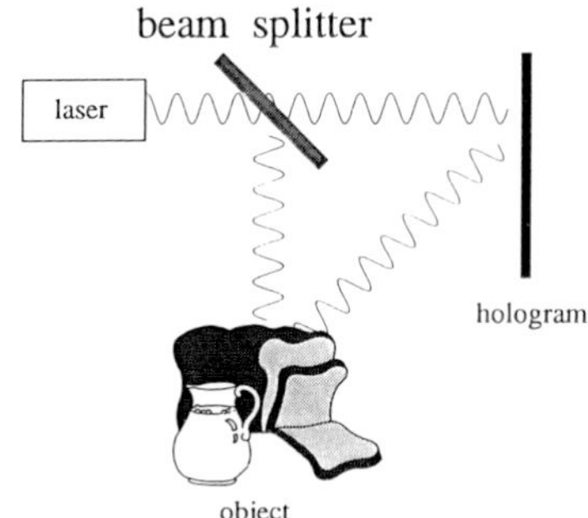

Figure 4. The principle of holography: a coherent beam of radiation, a laser in this instance, is split into two beams: one travels directly to the screen, the other illuminates the object.

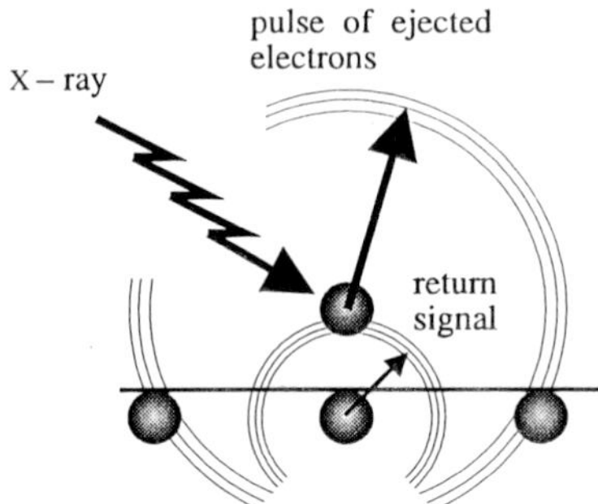

Figure 5. An electron-wave ejected from an adsorbed atom by an X-ray divides itself between a component escaping directly from the surface and a component that first scatters from atoms in the layer below, making an analogy with the holography experiment shown in figure 4.

Having drawn the analogy with holography, Szöke and Barton proposed to create an image of the surface by holographic reconstruction from the photoelectron diffraction pattern. That is a strong statement, and to what extent it can be made a reality is not clear.

There is one instance in which the idea has already been in use for some time. Suppose that figure 5 corresponds to a surface EXAFS experiment in which the interference pattern is measured at the emitting atom itself. The absorption cross-section as a function of energy is a one-dimensional hologram of the radial distribution function. It has long been recognized that Fourier transformation for EXAFS signals does yield a reasonable radial distribution function provided that one does not look beyond the first few shells, and that not too great an accuracy is required.

A clever extension of the one-dimensional (1D) hologram concept has been made by Wang *et al.* (1991). They observed that nearly all atoms have a strong peak in their scattering factor in the backwards direction. Hence diffracted intensity at a given angle of emission is dominated by direct emission plus any scattering from atoms immediately behind the emitter. They neglect all other contributions. Measuring the intensity as a function of energy at a given angle of emission gives a 1D hologram along a line immediately behind the emitter, in line of sight to the detector. Wang *et al.* convincingly make their case by Fourier transforming their 1D holograms taken at various angles of emission (figure 6) and show that they can

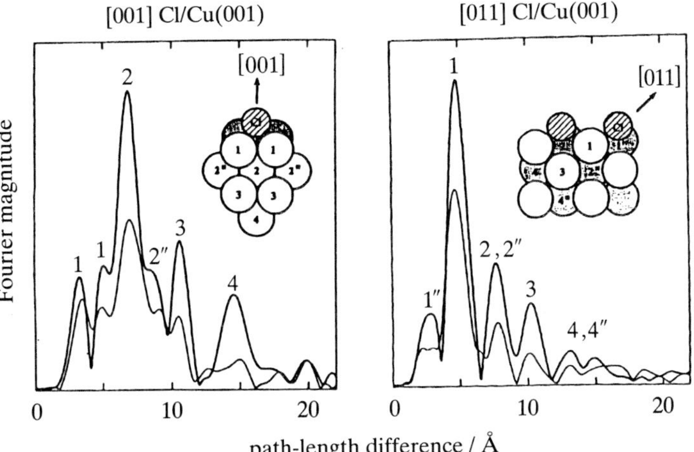

Figure 6. Fourier spectra for the C(2 × 2) Cl structure on Cu (100), after Wang *et al.* (1991). The data for this figure were taken with electrons in the [100] direction (left-hand figure) and [110] direction (right-hand figure). Data were taken at 110 K and 300 K, and each numbered peak is associated with a scattering path-length difference for a numbered atom in the inset. Note how changing the emission direction emphasizes those atoms behind the line of sight to the detector.

emphasize the signal from one or other of the surface atoms by aligning them with the back-scattering direction.

5. Conclusions

Quantitative surface crystallography is currently producing many new results on a diverse set of systems. The simplest of these can easily be handled by long-established techniques, but the more complex ones require help from recent advances. Such is the appetite of surface science for ever more information on the location of surface atoms that there is strong pressure for yet more advances in our ability to interpret data. Some of the more radical ideas such as electron surface holography are certainly in the developmental stage, but offer future possibilities for flexible on-line interpretation of diffraction data.

References

Andersson, S. & Pendry, J. B. 1980 *J. Phys.* C **13**, 3547.

Barton, J. 1988 *Phys. Rev. Lett.* **61**, 1356.

Binnig, G. & Rohrer, H. 1982 *Helv. phys. Acta* **55**, 726.

Fink, H.-W. 1986 *IBM J. Res. Develop.* **30**, 460.

Heinz, K. & Müller, K. 1982 In *Structural studies of surfaces*. Berlin: Springer.

MacLaren, J. M., Pendry, J. B., Rous, P. J., Saldin, D. K., Somorjai, G. A., Van Hove, M. A. & Vvedensky, D. D. 1987 *Surface crystallographic information service*. Dordrecht: Reidel.

Müller, E. W. & Tsong, T. T. 1969 *Field ion microscopy, principles and applications*. New York: Elsevier.

Pendry, J. B. 1974 *Low energy electron diffraction*. London: Academic.

Pendry, J. B. & Heinz, K. 1990 *Surf. Sci.* **230**, 137.

Rous, P. J. 1992 *Prog. surf. sci.* **39**, 3.

Szöke, A. 1986 In *Short wavelength coherent radiation: generation and applications* (ed. D. T. Attwood & J. Boker), AIP Conference Proceedings number 147. New York: AIP.

Van Hove, M. A. & Tong, S. Y. 1979 *Surface crystallography by LEED*. Berlin: Springer.
Wander, A., Van Hove, M. A. & Somorjai, G. A. 1991 *Phys. Rev. Lett.* **67**, 626.
Wander, A., Pendry, J. B. & Van Hove, M. A. 1992 *Phys. Rev. Lett.* (Submitted.)
Wang, L.-Q., Schach von Wittenau, A. E., Ji, Z. G., Wang, L. S., Huang, Z. Q. & Shirley, D. A. 1991 *Phys. Rev.* B **44**, 1292.

Discussion

A. M. Stoneham (*Harwell Laboratory, Didcot, U.K.*): Do diffraction methods give any information on properties of surfaces as well as their structure? Is it that direct real space methods (e.g. STM), even if cruder, are preferred on the issue of mechanisms?

J. B. Pendry: The STM gives information on a larger length scale than the crystallographic, which can be equally relevant to mechanisms.

Classical and quantum simulation of the surface properties of α-Al_2O_3

By W. C. Mackrodt

ICI Chemicals and Polymers Ltd, PO Box 8, The Heath, Runcorn, Cheshire WA7 4QD, U.K.

Classical simulations are described of the fully relaxed surface lattice structures of the five lowest-index planes of α-Al_2O_3 and the resulting crystal morphology. The surface coverage by yttrium and magnesium as a function of temperature is evaluated on the basis of a non-Arrhenius isotherm and calculated heats of surface segregation. The calculated morphology, surface coverages and heats of segregation are compared with experiment. A quantum simulation of a relaxed {0001} surface is presented and the surface structure and energy compared with the classical results. Estimates are made of the adsorption energy of HF at the {0001} surface.

1. Introduction

Oxide surfaces play an important role in a wide range of processes such as catalysis, corrosion, crystal growth, electrolysis and high-temperature superconductivity. However, despite the many advances in surface-sensitive analytic techniques that have occurred over the past decade or so, knowledge of the properties of even the simplest oxide surface remains sparse. For example, reference to the compilation of surface structures by MacLaren *et al.* (1987) reveals that of more than 260 structures cited only ten are for oxides, and then for only six different materials. Experimental surface energies are even more scarce. As elsewhere in solid-state physics and chemistry, computer-based theoretical methods offer a solution to this paucity of information that is increasingly both cost-effective and reliable. Accordingly, to illustrate what is now computationally feasible this paper describes classical and quantum simulations, as these computer-intensive methods are now frequently called, of the surface properties of α-Al_2O_3. α-alumina has been chosen for detailed discussion for a number of reasons: it is a paradigm ceramic with a long history of detailed investigation, though not of its surface properties; classical simulations predict appreciable surface relaxations (Mackrodt 1987), which could be verified directly by techniques such as quantitative I/V low-energy electron diffraction (LEED) and surface X-ray crystallography; there is good data on cation impurity segregation which should be amenable to theoretical interpretation; finally direct comparisons can now be made between classical and quantum simulations.

2. Classical simulations

(*a*) *Theory*

Classical simulations are based on the notion that the total internal potential energy of a system, E, can be written in the form

Phil. Trans. R. Soc. Lond. A (1992) **341**, 301–312
© 1992 The Royal Society
Printed in Great Britain

$$E = \sum_{i,j}{}' e_{ij} + \sum_{i,j,k}{}'' e_{ijk} + \dots, \tag{2.1}$$

in which e_{ij}, e_{ijk} etc are two-, three-, ... n-body terms acting between the constituent particles. Equation (2.1) can be rewritten as

$$E = \sum_{i,j}{}' e_{ij} \{1 + \sum_{k}{}'' e_{ijk}/e_{ij} + \dots\}, \tag{2.2}$$

$$E = \sum_{i,j}{}' \tilde{e}_{ij}, \tag{2.3}$$

with 'effective' two-body potentials, $\tilde{e}_{ij}$, defined as

$$\tilde{e}_{ij} = e_{ij} \{1 + \sum_{k}{}'' e_{ijk}/e_{ij} + \dots\}. \tag{2.4}$$

$\tilde{e}_{ij}$ can be obtained either by fitting analytic forms to known (bulk) properties such as the lattice structure, elastic and dielectric constants and phonon frequencies or by direct calculation using density-functional or Hartree–Fock methods (Catlow *et al.* 1982). All the classical simulations reported here are based on density-functional potentials (Mackrodt & Stewart 1979; Allan *et al.* 1991) which incorporate the shell-model of Dick & Overhauser (1958) to allow for electronic polarization effects (Catlow *et al.* 1982).

The presence of a free surface or any other discontinuity in a crystal, including point defects, leads to a relaxation or distortion of the lattice from the perfect bulk structure. In principle, the relaxation, Q_s, at a surface, s, and the corresponding surface energy, E_s, can be determined from the perfect lattice Green function, G, and the change in the total potential at s, δV_s. For practical purposes, however, the structure (and energy) of the relaxed lattice in the static approximation is more conveniently determined from the condition that the (defective) system is in mechanical equilibrium, i.e.

$$\partial E/\partial X_i = 0, \tag{2.5}$$

where E is the total energy and X_i are all the variables that define the structure. In the case of surfaces, the bulk termination is the most useful reference configuration so that X_i comprise the displacements, Q_s, of the atomic coordinates (and changes in the shell-model polarization) from the bulk positions of atoms/ions in the topmost surface planes. The generality of this approach is such that pure and impure surfaces can be treated in an identical way, which has led to many of the more recent studies on impurity segregation, including those discussed below.

The predominant relaxation is perpendicular to the surface, resulting in a dilation of the surface planes (change in the average separation between them) and rumpling (differential relaxation of unlike atoms/ions in a given plane). Relaxation parallel to the surface is also possible, as shown by Tasker (1984) for the $\{10\bar{1}0\}$ surface of α-Al_2O_3. For most ionic systems, including oxides, relaxation normally decreases fairly rapidly away from the surface, so that in general there is seldom need to relax more than the top five or six planes for simple low-index surfaces. High-index stepped surfaces, on the other hand, frequently require the relaxation of many more planes. The most useful and flexible computer code for calculating the surface structure and energy of ionic materials based on equation (2.1) is the MIDAS code written by Tasker (1978) and the surfaces of a wide variety of ionic crystals have been studied using this.

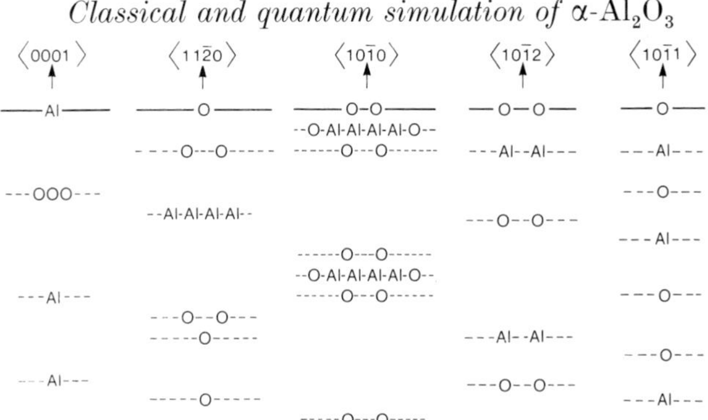

Figure 1. Stacking sequences of the five lowest-index surfaces of α-Al_2O_3.

Table 1. *Classical surface energies* ($J\ m^{-2}$) *of* α-Al_2O_3

surface	unrelaxed	relaxed
{0001}	5.95	2.03
{1010}	6.46	2.23
{1012}	3.63	2.29
{1120}	4.37	2.50
{1011}	5.58	2.52

(*b*) *Surface structure*

The surface of α-Al_2O_3 is dominated by the five lowest-index planes, the stacking sequences of which are shown in figure 1. The energies of the unrelaxed surfaces based on electron-gas potentials (Mackrodt 1987) are given in table 1. Thus the relative stability of these surfaces is

$$\{10\bar{1}2\} < \{11\bar{2}0\} < \{10\bar{1}1\} < \{0001\} < \{10\bar{1}0\}$$

in hexagonal notation. Allowing the surfaces to relax to mechanical equilibrium, reduces the surface energies appreciably and changes the order to

$$\{0001\} < \{10\bar{1}0\} \approx \{10\bar{1}2\} < \{11\bar{2}0\} \approx \{10\bar{1}1\},$$

in which the basal surface is now lowest in energy. The corresponding energies are given in table 1. Thus lattice relaxation has three distinct effects. First, it modifies the atomic structure of the surfaces, with changes in lattice spacing of up to 60%, as shown in figure 2 for the {0001} surface. Second, it lowers the individual surface energies by up to a factor of three. Third, it re-orders the relative stability of the surfaces. While there is no direct confirmation as yet for these effects, there is supportive evidence. An approximate value of $0.9\ J\ m^{-2}$ has been reported for the surface energy of α-Al_2O_3 at 1850 °C (Kingery *et al.* 1976), which presumably represents an average surface energy at this temperature. Following Tasker (1984), the surface energy at a temperature T, $\gamma(T)$, is written in the approximate form

$$\gamma(T) \approx \gamma(0) + \Gamma T, \tag{2.6}$$

from which $\gamma(0)$, the average value at 0 K can be deduced. If, in the absence of a value of Γ for α-Al_2O_3 and as a further approximation, the reported coefficient of $-6.3\times10^{-4}\ J\ m^{-2}\ K^{-1}$ for UO_2 (Hodkin & Nicholas 1973) is used, a value of $2.2\ J\ m^{-2}$ is obtained for $\gamma(0)$. This compares with an unrelaxed average surface

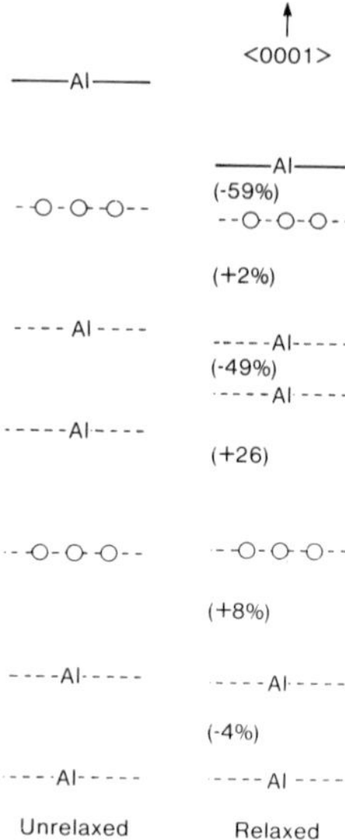

Figure 2. Comparison of the unrelaxed and classically relaxed stacking structure of the {0001} surface of α-Al_2O_3. (Figures in brackets are the percentage changes in the interplanar spacings.)

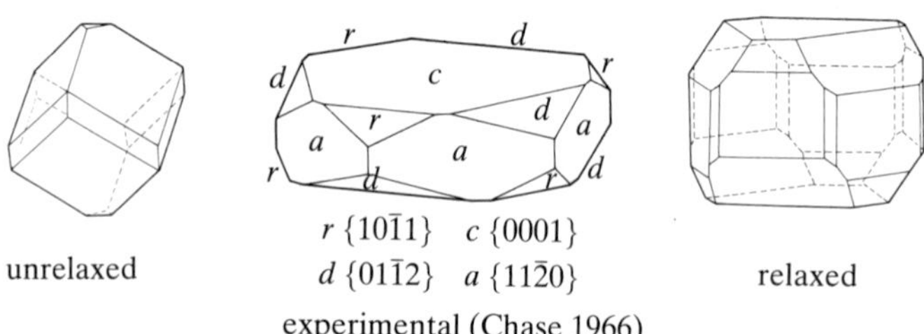

Figure 3. Unrelaxed and relaxed morphologies of α-Al_2O_3 compared with experiment. (Courtesy of the *J. Am. Ceram. Soc.*)

energy of 5.2 J m^{-2} and a value of 2.3 J m^{-2} for fully relaxed surfaces. The relative stability of the individual surfaces may be assessed by invoking the Wulff relation

$$\gamma_i/h_i = \gamma_j/h_j \quad \text{for all} \quad i,j, \tag{2.7}$$

to predict the single-crystal morphology, in which γ_i is the surface energy of surface i and h_i the perpendicular distance of i to a common centre. Figure 3 shows the predicted morphologies for crystals with unrelaxed and relaxed surfaces (Mackrodt *et al.* 1987) compared with an experimental morphology reported by Chase (1966). There is evidently overall agreement between the relaxed and experimental morphologies, particularly in respect of the predominant {0001} face which is barely present in the unrelaxed structure.

(c) {0001} *surface dynamics*

It is possible to go beyond the static approximation and examine the dynamics of alumina surfaces based Cochran's (1977) extension of the lattice dynamics originally due to Born & von Karman (Allan & Mackrodt 1989). Taking the {0001} surface as an example, it is represented by a slab consisting of six fully relaxed stoichiometric units of alumina, i.e. 18 atomic layers. Phonon frequencies are obtained from the

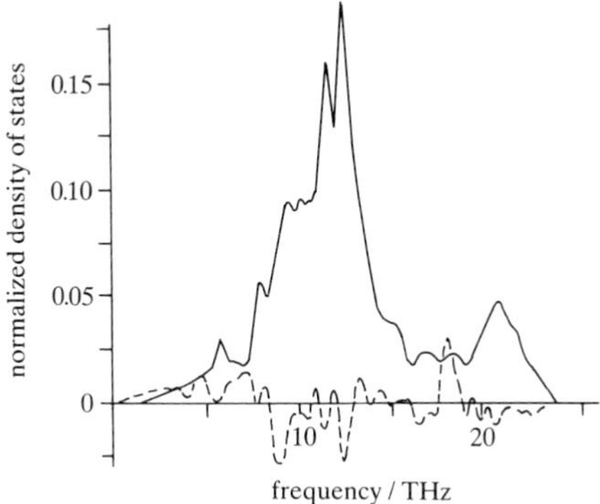

Figure 4. Calculated bulk (——) and surface-excess (– – – – –) phonon densities of states of the {0001} surface of α-Al_2O_3.

eigenvalues of the dynamical matrix which contains both coulombic and non-coulombic contributions. They are extremely sensitive to lattice stability and for lattices which are not at equilibrium, such as the unrelaxed slab, imaginary frequencies are found. The surface-excess phonon density of states is shown in figure 4, together with that for the bulk. The strongest of the surface peaks is predicted at 18 THz, with weaker intensities at approximately 4, 6 and 13 THz. As yet there appears to be no experimental data with which to compare these results.

(*d*) *Impurity segregation*

Ceramics in general, and oxides in particular, can seldom be produced in a completely pure state. Indeed, the presence of impurities is essential to many important processes involving oxides, including catalysis, corrosion and perhaps most notably high T_c superconductivity. Since boundaries of whatever sort provide a different environment from that of the bulk, there is always a free-energy gradient associated with impurities (and other point defects) which drives them to or from a boundary. What simulations can often provide, especially in those cases where experiments are extremely difficult, is information such as changes in surface structure, including the formation of second coherent phases, heats of segregation and equilibrium coverages as a function of temperature and bulk impurity concentration. At thermal equilibrium, the distribution of impurities at a surface is most conveniently described in terms of a segregation isotherm, which is frequently taken to be of the Langmuir or Arrhenius form,

$$[\text{surface}]/[\text{bulk}] = A\exp(-h/k_B T), \tag{2.8}$$

in which [surface] and [bulk] are the surface and bulk concentrations of impurity and h, the heat of segregation, which is assumed to be independent of coverage. However, other than at extremely low coverage this is seldom the case and Mackrodt & Tasker (1989) have derived a more general expression,

$$x_s = x_b \exp\{[s(x_s)+x_s(x_s+1)(ds(x_s)/dx_s)]/k_B\} \times \exp\{-[h(x_s)+x_s(x_s+1)(dh(x_s)/dx_s)]/k_B T\} \tag{2.9}$$

in which x_s is the ratio of impurity:host ions at the surface and not a concentration. $s(x_s)$ and $h(x_s)$ are the *coverage-dependent* entropy and enthalpy of segregation respectively. They are defined as the differences in the entropy and enthalpy for an isolated atom in the bulk and at the surface. Since the entropic contribution is temperature independent, equation (2.9) can be rewritten as

$$x_s = x_b A' \exp\{-[h(x_s)+x_s(x_s+1)\,(\mathrm{d}h(x_s)/\mathrm{d}x_s)]/k_B T\} \tag{2.10}$$

for direct comparison with equation (2.8). An important difference between equations (2.8) and (2.10) is that whereas H is the slope of ln ([surface]) against $1/k_B T$, $h(x_s)$ is not equal to the slope of $\ln(x_s)$ against $1/k_B T$ and cannot be obtained from it. Fortunately, $h(x_s)$ can be calculated from lattice simulations. However, before considering two specific examples of impurity segregation, it is worth considering the interpretation of those cases in which there is a linear relation between $\ln(x_s)$ and $1/T$.

Rewriting equation (2.6) as

$$\ln(x_s) = \ln(x_b A') - [h(x_s)+x_s(x_s+1)\,(\mathrm{d}h/\mathrm{d}x_s)]/k_B T\} \tag{2.11}$$

$$= B - H/k_B T, \tag{2.12}$$

solutions are sought to the equation

$$h(x_s)+x_s(x_s+1)\,(\mathrm{d}h(x_s)/\mathrm{d}x_s) = H, \tag{2.13}$$

where H is a constant.

The equation

$$h(x_s) = H = h \tag{2.14}$$

is clearly the solution when $\mathrm{d}h(x_s)/\mathrm{d}x_s = 0$, so that one interpretation of a linear relation between $\ln(x_s)$ and $1/T$ is that embodied in the Langmuir–Arrhenius isotherm. The general solution, however, is

$$h(x_s) = H - C(1+1/x_s), \tag{2.15}$$

in which C is an undetermined constant. Thus, systems for which $h(x_s) \propto 1/x_s$ will also exhibit Langmuir–Arrhenius behaviour with the possibility that $h(x_s)$ and H might be quite different.

Y : Al_2O_3

Yttrium is isovalent with aluminium so that it is a convenient impurity for theoretical study since, to a good approximation, other point defects can be neglected. It is also one of the few cation impurities in alumina for which there is segregation data (McCune *et al.* 1986). In the temperature range 1700 °C to 1900 °C the reported surface coverage of polycrystalline samples is *ca.* 13% with a heat of segregation of 44 kJ mol^{-1}, although the data are complicated by extensive surface and intergranular precipitation of yttrium aluminium garnet (YAG). Atomistic lattice simulations (Mackrodt 1987) find positive heats of segregation for all but the $\{0001\}$ surface and highly non-Arrhenius behaviour, as shown in figure 5 for the $\{10\bar{1}2\}$ and $\{11\bar{2}0\}$ surfaces. The values of $h(x_s)$ are typically in the range 100–400 kJ mol^{-1}, which is considerably greater than the reported value. The apparent discrepancy between theory and experiment is resolved by comparing the surface coverages. Figure 6 shows that for the $\{10\bar{1}2\}$ and $\{1120\}$ surfaces, for example, the calculated coverages are very nearly linear in the range 1700–1900 °C, with slopes of -13 and -35 kJ mol^{-1} respectively. These are much more in line with the experimental slope of -44 kJ mol^{-1}, which emphasizes quite clearly the difference between the Arrhenius heat, i.e. the slope of $\ln(x_s/x_b)$ against $1/T$, and

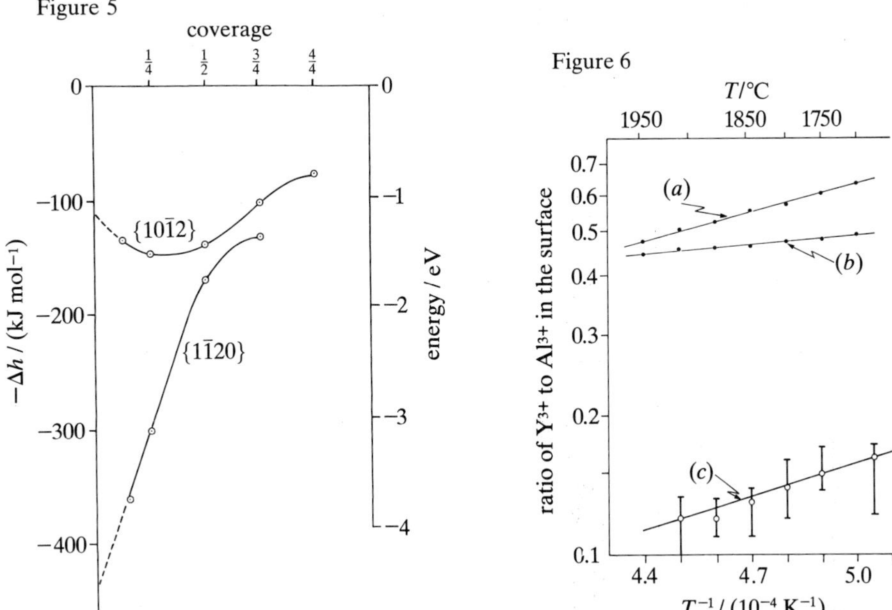

Figure 5. Calculated heat of segregation of yttrium at the $\{10\bar{1}2\}$ and $\{11\bar{2}0\}$ surfaces of α-Al_2O_3.
Figure 6. Calculated (●) and experimental (○) equilibrium coverages of α-Al_2O_3 by yttrium (160 ppm) as a function of temperature. (*a*) $\{10\bar{1}2\}$ surface, $H = -41$ kJ mol^{-1}. (*b*) $\{11\bar{2}0\}$ surface, $H = -13$ kJ mol^{-1}. (*c*) Φ experimental coverage of polycrystalline α-Al_2O_3 (McCune *et al.* 1986), $H = -44$ kJ mol^{-1}.

what might be called the atomic the heat of segregation, $h(x_s)$. There is a discrepancy between the magnitude of the calculated and observed coverages, for which there are essentially three possible reasons. The first is that the very few calculations of the entropy of segregation that have been carried out suggest that A' in equation (2.10) is less than unity. Thus for Ca at {100}MgO the entropy reported by Masri *et al.* (1986) leads to value of 0.9. The second is that the experimental coverage is an average for a finite depth into the crystal which may differ from the surface coverage. The third is that the reported data is for polycrystalline material which is likely to contain a substantial proportion of {0001} surface. Calculations predict negligible segregration to this surface, which, if correct, implies a greater coverage of the other surfaces. Together these suggest closer agreement between theory and experiment than might seem to be case.

Mg: Al_2O_3

The treatment of aliovalent impurities generally is complicated by the need to include the charge compensating defects. In the case of Mg in Al_2O_3 it is believed that oxygen vacancies are the neutralizing species and on this basis the heat of segregation of Mg to the {0001} surface of alumina as the neutral complex, $\{Mg_{Al}$–V_O–$Mg_{Al}\}$, has been calculated (Mackrodt & Tasker 1986): it is shown in figure 7. From zero to about $\frac{1}{4}$ coverage $h(x_s)$ is nearly constant. Beyond this it falls sharply to a minimum at about $\frac{1}{3}$ coverage, after which it rises to a maximum at around $\frac{3}{4}$ coverage. At the minimum value of $h(x_s)$, the ratio of Al^{3+} to Mg^{2+} ions is exactly that of spinel, $MgAl_2O_4$, which suggests the formation of a second coherent phase with this composition at the surface. The corresponding equilibrium coverage is shown in

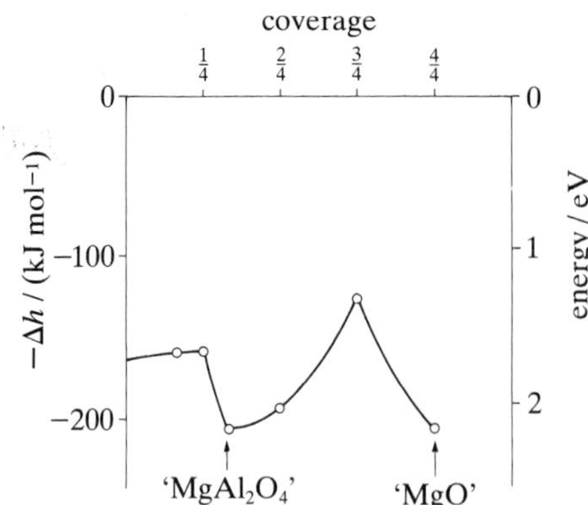

Figure 7. Calculated heat of segregation of magnesium at the {0001} surface of α-Al_2O_3.

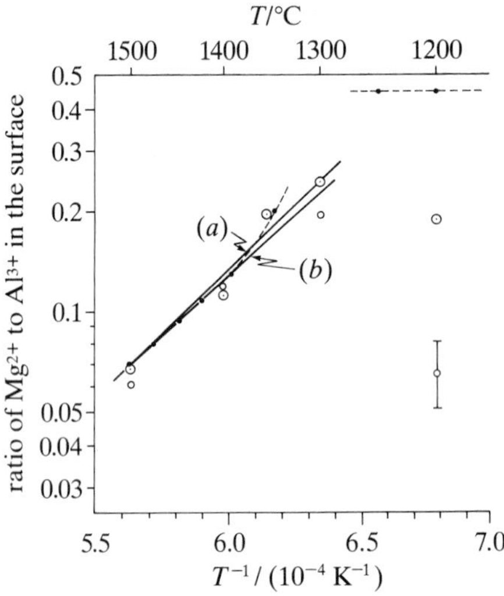

Figure 8. Calculated (●) and experimental (○, 1 $h\nu$; ⊙, 10 $h\nu$ (Baik *et al.* 1985)) equilibrium coverages of the {0001} surface of α-Al_2O_3 by magnesium (40 ppm) as a function of temperature. (a) Experimental, $H = -146$ kJ mol^{-1}; (b) calculated $H = -158$ kJ mol^{-1}.

figure 8, together with data reported by Baik *et al.* (1985). Theory and experiment both find Arrhenius-like behaviour up to *ca.* 1370 °C, with slopes that agree to approximately 10 kJ mol^{-1}. Between about 1350 °C and 1250 °C there is a sharp increase in the calculated coverage corresponding to the minimum in $h(x_s)$ and it is not unreasonable to relate the predicted spinel-like second phase to the putative cubic overlayer suggested by the changes in the observed LEED patterns (Baik *et al.* 1985).

3. Quantum simulations

Recent advances in computer technology have made possible a start to the *ab initio* quantum simulation of oxide and other surfaces. Inevitably the size of quantum simulations will be considerably less than classical analogues. However, they will enable direct comparisons to be made in certain cases and beyond this, test some of the assumptions commonly made in classical treatments. For example, classical simulations assume that the charge and polarizability of surface ions and

Table 2. *Quantum simulation of the {0001} surface of alumina*

	ionic charges (Bulk: Al, +2.25; O, −1.50.)	
layer	unrelaxed	relaxed
Al (s)	+1.84	+2.08
O (s−1)	−1.31	−1.39
Al (s−2)	+2.07	+2.10
Al (s−3)	+2.27	+2.26
O (s−4)	−1.50	−1.52

	surface energies (J m^{-2})	
	classical	quantum
unrelaxed	5.95	3.20
relaxed	2.03	2.00

	lattice relaxation	
spacing	classical	quantum
s_1	−59%	−49%
s_2	+2%	−5%
s_3	−49%	−8%

the resulting interionic potentials, are essentially that of the bulk. Quantum simulations will allow this to be tested. The quantum methodology used in the examples reported here is the *ab initio* localized orbital Hartree–Fock method (Pisani *et al.* 1988) discussed elsewhere in this volume (Dovesi *et al.* 1992). It has been used to calculate the bulk electronic structure of α-Al_2O_3 (Salasco *et al.* 1991) and the surface features of MgO (Causa *et al.* 1986). The atomic bases used in the simulations reported here (Gale *et al.* 1992) are modifications of the standard 6-21G Pople functions (Binkley *et al.* 1980; Gordon *et al.* 1982). They comprise a 6-21G set for Al and an 8-51G set for O (Dovesi 1985) with the outer exponents of both sets optimized for the experimental bulk structure. This leads to a more ionic structure, than that reported previously, with a Mulliken charge of +2.25 for Al as compared with +1.775 and +2.024 for basis sets with and without Al d-functions (Salasco *et al.* 1991).

(*a*) *{0001} surface structure*

The model for the {0001} surface examined recently by Gale *et al.* (1992) is an infinite two-dimensional slab consisting of up to three stoichiometric units of Al_2O_3 per unit cell, i.e. nine atomic layers perpendicular to the $\langle 0001 \rangle$ direction. The unrelaxed surface structure is that of the experimental bulk termination. The Mulliken charge of the surface Al is +1.84 and that of the underlying O is −1.31: these compare with bulk values of +2.25 and −1.50. However, as table 2 shows, the values for third Al layer and second O layer are very close to the bulk charges. The surface energy of the unrelaxed slab is 3.30 J m^{-2} compared with the classical value based on electron-gas potentials of 5.95 J m^{-2} (Mackrodt 1987). Relaxation of the outer three atomic layers, two Al and one O, leads to an increase in the surface Mulliken charges to +2.08 and −1.39 and a reduction in the surface energy to 2.00 J m^{-2} which is close to the classical value of 2.03 J m^{-2} (Mackrodt 1987). There

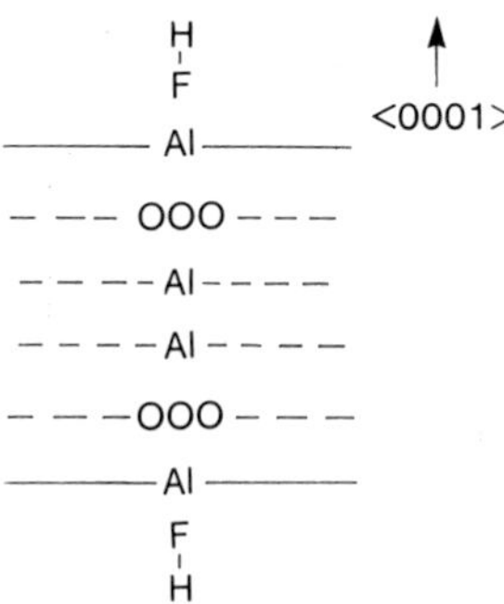

Figure 9. Slab configuration for the quantum simulation of HF adsorbed at the {0001} surface of α-Al_2O_3.

is also good agreement between the surface structures derived from the two methodologies. Classical simulations predict a contraction of the outermost interlayer spacing of 59% compared with a 49% contraction by quantum simulation. Both predict the next spacing to be close to the bulk value with deviations of +2% and −8% respectively. However, as shown in table 2, there is a difference for the third layer, but this might be due to the fixed inner-most spacing in the quantum simulation.

(*b*) *Sorption of* HF *and* HCl

The use of localized orbitals in the *ab initio* crystal Hartree–Fock approach of Pisani *et al.* (1988) leads to its natural extension to adsorption studies at crystal surfaces (Dovesi *et al.* 1987). Previous calculations of the adsorption energy of CO at {100} MgO found good agreement with experiment for $\frac{1}{2}$ and $\frac{1}{4}$ surface coverage (Dovesi *et al.* 1987). Gale *et al.* (1992) have studied the sorption of hydrogen fluoride at the basal surface of alumina, a process which is relevant to the catalytic fluorination of chlorocarbons (Hedge & Barteau 1989). Their simplest model consists of a fully-relaxed, infinite, two-dimensional slab containing six atomic layers interacting with a single molecule of HF per surface unit cell at both surfaces, as shown in figure 9. It corresponds to a surface coverage of $5.1\times10^{-}$ molecules Å^{-2}†. A standard 6-21G basis set was used for HF which leads to an equilibrium bond length of 0.939 Å and dipole moment of 2.15 D, compared with experimental values of 0.917 Å (Kuipers *et al.* 1956) and 1.83 D (Muenter & Klemperer 1970). For the configuration shown in figure 9, the minimum-energy Al–F distance was found to be 1.93 Å, which suggests an essentially long-range interaction between the HF dipole and the electrostatic potential of the surface. At this separation there is an increase in the HF bond length by 0.5% to 0.943 Å and an expansion of the outer-most lattice spacing by about 6%. The binding energy is calculated to be 61.2 kJ mol^{-1} molecule^{-1} with a charge transfer of *ca.* 0.1 e to the surface. Calculations of two-dimensional arrays of HF in the adsorbed configuration indicate that there is a negligible interaction between the HF layers at the two surfaces and a lateral repulsive interaction within each layer of about 20 kJ mol^{-1} molecule^{-1}. This compares with an energy of *ca.* 4 kJ mol^{-1} for the nearest neighbour repulsion. The estimate of the lateral interaction in the adsorbed state is probably low since it does not include the effect of the charge transfer to the surface. However, taken as a first approximation, it leads to an estimate of 81 kJ mol^{-1} for the net binding energy of

† 1 Å = 10^{-10} m = 10^{-1} nm.

an HF molecule and the {0001} surface of $-Al_2O_3$. Further calculations of different surface configurations and coverage dependence, which are in progress (Gale *et al.* 1992), should lead to a more reliable value for the binding energy and an estimate of the coverage-dependent heat of adsorption.

References

Allan, N. L. & Mackrodt, W. C. 1989 Calculated surface phonon densities of states of ionic oxides and fluorides. *J. Phys. Condensed Matter* **1**, 189–190.

Allan, N. L., Cooper, D. L. & Mackrodt, W. C. 1991 The practical calculation of interionic potentials in solids using electron gas theory. *Molecular Simulation* **4**, 269–283.

Baik, S., Fowler, D. E., Blakely, J. M. & Raj, R. 1985 Calcium segragation to MgO and α-Al_2O_3 surfaces. *J. Am. ceram. Soc.* **68**, 281–286.

Binkley, J. S., Pople, J. A. & Hehre, W. J. 1980 Self-consistent molecular orbital methods. 21. Small split-valence basis sets for first-row elements. *J. Am. chem. Soc.* **102**, 939–947.

Catlow, C. R. A., Dixon, M. & Mackrodt, W. C. 1982 Interionic potentials in ionic solids. In *Computer simulation of solids* (ed. C. R. A. Catlow & W. C. Mackrodt). Berlin: Springer-Verlag.

Chase, A. B. 1966 Habit modification of corundum crystal grown from molten PbF_2–Bi_2O_3. *J. Am. ceram. Soc.* **49**, 233–236.

Causa, M., Dovesi, R., Pisani, C. & Roetti, C. 1986 *Ab initio* Hartree–Fock structure of the MgO (100) surface. *Surf. Sci.* **175**, 551–560.

Cochran, W. 1977 Lattice dynamics of ionic and covalent crystals. *Crit. Rev. Sol. St. Sci.* **2**, 1–44.

Dick, B. G. & Overhauser, A. W. 1958 Theory of dielectric constants of alkali halide crystals. *Phys. Rev.* **112**, 90.

Dovesi, R. 1985 *Ab initio* Hartree–Fock extended basis set calculation of the electronic structure of crystalline lithium oxide. *Solid State Commun.* **54**, 183–185.

Dovesi, R., Orlando, R., Ricca, F. & Roetti, C. 1987 CO adsorption on MgO crystals: Hartree–Fock calculations for regular adlayers on a (001) lattice plane. *Surf. Sci.* **186**, 267–278.

Dovesi, R., Roetti, C., Freyria-Fava, C., Aprà, E., Saunders, V. R. & Harrison, N. M. 1992 *Phil. Trans. R. Soc. Lond.* A **341**, 203–210.

Gale, J. D., Catlow, C. R. C. & Mackrodt, W. C. 1992 Periodic *ab initio* determination of intermolecular potentials for aluming. *Model. Simulat. Mater. Sci. Engng* (In the press.)

Gordon, M. S., Binkley, J. S., Pople, J. A., Pietro, W. J. & Hehre, W. J. 1982 Self-consistent molecular orbital methods. 22. Small split-valence basis sets for second-row elements. *J. Am. chem. Soc.* **104**, 2797–2803.

Hedge, R. I. & Barteau, M. A. 1989 Preparation, characterisation and activity of fluorinated aluminas for halogen exchange. *J. Catalysis* **120**, 387–400.

Hodkin, E. N. & Nicholas, M. G. 1973 Surface and interfacial properties of stoichiometric uranium dioxide. *J. nucl. Mater.* **47**, 23–30.

Kingery, W. D., Bowen, H. K. & Uhlmann, D. R. 1976 *Introduction to ceramics*, 2nd edn. Wiley.

Kuipers, G. A., Smith, D. F. & Nielsen, A. N. 1956 Infrared spectrum of hydrogen fluoride. *J. chem. Phys.* **25**, 275–279.

Mackrodt, W. C. 1987 The calculated equilibrium segregation of Fe^{3+}, Y^{3+} and La^{3+} at the low-index surfaces of α-Al_2O_3. In *Advances in ceramics*, vol. 23 (ed. C. R. A. Catlow & W. C. Mackrodt), pp. 293–306.

Mackrodt, W. C. & Stewart, R. F. 1979 Defect properties of ionic solids: II Point defect energies based on modified electron-gas potentials. *J. Phys.* C **12**, 431–449.

Mackrodt, W. C. & Tasker, P. W. 1986 Calculated impurity segregation at the surfaces of α-Al_2O_3. *Mater. Res. Soc. Symp. Proc.* **60**, 291–298.

Mackrodt, W. C. & Tasker, P. W. 1989 Segregation isotherms at the surfaces of oxides. *J. Am. ceram. Soc.* **72**, 1576–1583.

Mackrodt, W. C., Davey, R. J., Black, S. W. & Docherty, R. 1987 The morphology of α-Al_2O_3 and α-Fe_2O_3: the importance of surface relaxation. *J. Crystal Growth* **80**, 441–446.

MacLaren, J. M., Pendry, J. B., Rous, P. J., Saldin, D. K., Somorjai, G. A., Van Hove, M. A. & Uvedensky, D. D. 1987 *Surface crystallographic information service: a handbook of surface structures.* Dordrecht: Reidel.

Masri, P., Tasker, P. W., Hoare, J. P. & Harding, J. H. 1986 Entropy and segregation of impurity cations at the surface of an ionic crystal: MgO (001): Ca^{2+}. *Surf. Sci.* **173**, 439–454.

McCune, R. C., Donean, I. N. T. & Ku, R. C. 1986 Yttrium segragation and YAG precipitation at the surfaces of Y-doped α-Al_2O_3. *J. Am. ceram. Soc.* **69**, C196-C199.

Muenter, J. S. & Klemperer, W. 1970 Hyperfine structure constants of HF and DF. *J. chem. Phys.* **52**, 6033–6037.

Pisani, C., Dovesi, R. & Roetti, C. 1988 Hartree–Fock *ab initio* treatment of crystalline systems. *Lecture Notes in Chemistry*, vol. 48. Springer-Verlag.

Salasco, L., Dovesi, R., Orlando, R., Causa, M. & Saunders, V. R. 1991 A periodic *ab initio* extended basis set study of α-Al_2O_3. *Molec. Phys.* **72**, 267–277.

Tasker, P. W. 1978 *A guide to MIDAS.* A.E.R.E. Report R.9130.

Tasker, P. W. 1984 Surfaces of magnesia and alumina. In *Advances in ceramics*, vol. 10 (ed. W. D. Kingery), pp. 176–189.

Discussion

Sir John Meurig Thomas (*The Royal Institution, London, U.K.*): Most oxides have hydroxylated surfaces; they seldom end with O^{2-} or M^{2+}. Is the degree of relaxation that you reported for all the prismatic and pyramidal faces (ending in O^{2-} in the idealized state) also seen when you repeat your calculation for α-Al_2O_3 terminating in OH groups?

W. C. Mackrodt: Calculations of the type you refer to have not been carried out for α-Al_2O_3, as far as I am aware. However, similar calculations for MgO indicate that hydroxylation does effect surface relaxation.

G. Ackland (*Department of Physics, The University of Edinburgh, U.K.*): The fact that the potential is the same at the bulk and at the surface means that the charge state of the O atom is irrelevant to the relaxation. The Hartree–Fock calculation seems primarily to create O^{2-} ions, and subsequent relaxation is mainly electrostatic and therefore well approximated by the shell model.

Atomistic modelling of diffusional phase transformations

By A. Cerezo[1], J. M. Hyde[1], M. K. Miller[2], S. C. Petts[1]†, R. P. Setna[1] and G. D. W. Smith[1]

[1] *Department of Materials, University of Oxford, Parks Road, Oxford OX1 3PH, U.K.*
[2] *Metals and Ceramics Division, Oak Ridge National Laboratory, Oak Ridge, Tennessee 37831-6376, U.S.A.*

A simple Monte Carlo model has been used to simulate diffusional phase transformations occurring in binary alloys during thermal ageing. The results of the simulation are compared directly with atomic-scale chemical information obtained from position-sensitive atom probe microanalysis. A simple pair potential model is found to give a good match with the ageing behaviour of spindodally decomposing iron–chromium alloys. Initial results on the modelling of nucleation and growth are presented, and compared with phase separation of copper–cobalt alloys.

1. Introduction

Conventional theories of diffusional phase transformations are based on continuum models, an approach which appears reasonable given the number of atoms in even a modest amount of material. However, advances in microscopy and microanalysis have extended the study of phase decompositions to the earliest stages, when composition fluctuations are on the scale of only a few atomic spacings. The properties of many technological alloys depend on variations in chemical composition at the nanometre scale. Structures of this size cannot be adequately represented using continuum representations of the composition, and this leads to a breakdown of conventional models in predicting the kinetics of the reactions. Understanding the early stages of phase decomposition is not only important at a fundamental level, but also for predicting the long-term stability of engineering alloys and in the design of new materials. This paper describes an attempt to provide a statistical, atomistic model of the ageing of materials, through the use of a Monte Carlo simulation. The modelling work is being performed in conjunction with an experimental programme of atomic-scale studies of the early stages of phase transformations. It is therefore possible to make direct comparisons between the results of the atomistic simulation and the atomic-scale chemical measurements.

2. Atom probe microanalysis

The technique of atom probe microanalysis (Müller *et al.* 1968) provides the chemical analysis of materials at the highest possible spatial resolution. (For a general review of the technique and its applications see, for example, Miller & Smith (1989).) In this instrument a specimen in the form of a sharp needle point is held in an ultra-high vacuum chamber and subjected to a high positive potential. The

† Present address: 65 Ferndale Road, New Milton, Hants BH25 5EX, U.K.

Phil. Trans. R. Soc. Lond. A (1992) **341**, 313–326
© 1992 The Royal Society
Printed in Great Britain

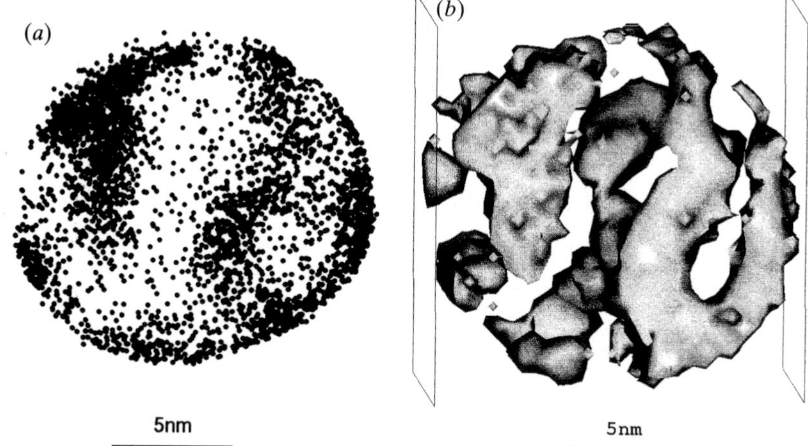

Figure 1. Position sensitive atom probe analysis of the ferrite phase in a CF3 duplex stainless steel (Fe, 21 % Cr, 9 % Ni, 0.4 % Mn, 0.7 % Si, 0.2 % Mo, 0.02 % C (all by mass)) aged for 10^4 hours at 400 °C. (*a*) Chromium atom map from a slice of material 2 nm deep, showing the clustering of chromium atoms into an interconnected structure during the thermal ageing. (*b*) Isosurface reconstruction of the microstructure from a region of material 15 nm in diameter and 10 nm deep. After a local sampling process, the surface is drawn through all points of composition 30 at. %, which represents the interface between chromium enriched and chromium depleted regions.

intense field is sufficient to ionize prominent atoms on the specimen surface, and the resulting ions travel along approximately radial paths away from the specimen point. This radial projection generates a magnification of approximately 10^6. A small aperture in the system selects ions from a region 1–2 nm in diameter on the specimen surface, allowing them to pass into a time-of-flight mass spectrometer. The elemental identification of single atoms removed from the surface gives a measurement of the local composition of the selected region, and continued removal gives a measurement of the variation of composition with depth, with single atomic layer depth resolution. A recent extension of this technique, the position sensitive atom probe (POSAP) (Cerezo *et al.* 1989) uses a position-sensitive detector to identify ions removed from a larger area (approximately 20 nm in diameter) on the specimen surface. From the position of arrival and identity of individual ions, an atomic map is built up of the distribution of elements on the specimen surface (figure 1*a*). Continued removal of material allows a full three-dimensional reconstruction of the atomic positions, and thus the variations in chemical composition (figure 1*b*). The POSAP analysis of the ferrite phase in a CF3 duplex stainless steel (Fe, 21 % Cr, 9 % Ni, 0.4 % Mn, 0.7 % Si, 0.2 % Mo, 0.02 % C (all by mass)), aged for 10^4 h at 400 °C is shown in figure 1. This example shows the analysis of a volume 15 nm in diameter and 10 nm in depth, showing the scale of the structures which are observable in the POSAP. Despite the significant thermal ageing of this material, the wavelength of the microstructure formed is less than 10 nm, or approximately 50 atom spacings.

3. Monte Carlo simulation

In the computer simulation, an alloy is represented by an Ising model with conserved order parameter. Atoms of type A or B (equivalent to spin-up or spin-down) are placed on a regular three-dimensional cubic lattice. Much of the simulation has been done on a simple cubic lattice but body-centred and face-centred lattices have also been used. Diffusional processes in the solid are simulated either by atom swaps, or by a random walk of a vacancy through the lattice. At any stage, the new configuration (after an atom–atom or atom–vacancy swap) is accepted or discarded according to the change in energy $\Delta E = E_{i+1} - E_i$. The probability of acceptance is given by the Metropolis function (Metropolis *et al.* 1953):

$$W(X_i \rightarrow X_{i+1}) = \begin{cases} \exp(-\Delta E/kT) & \text{if} \quad \Delta E > 0, \\ 1 & \text{if} \quad \Delta E \leqslant 0. \end{cases}$$

To date, atomic binding has been represented in the model using a simple pair potential, and no atom–vacancy binding energy has been included. Although the modelling of diffusion by atom swaps is simpler, this method has the disadvantage that it can only model symmetrical systems, since the energy change on any swap will always be a multiple of the energy parameter

$$\alpha = E_{\mathrm{aa}} + E_{\mathrm{bb}} - 2E_{\mathrm{ab}},$$

where E_{aa} is the binding energy between two A atoms, E_{bb} is that between two B atoms, and E_{ab} the binding between two unlike atoms. There is no method by which a difference between E_{aa} and E_{bb} can be represented in the simulation. However, an atom–vacancy swap is described in terms of two energy parameters:

$$\alpha_1 = E_{\mathrm{aa}} - E_{\mathrm{ab}} \quad \text{and} \quad \alpha_2 = E_{\mathrm{bb}} - E_{\mathrm{ab}}.$$

To make a comparison between the two simulation methods, the binding energies for the two cases must be made identical, that is $\alpha_1 = \alpha_2 = \frac{1}{2}\alpha$. Under these conditions, no difference is found between simulations using atom swapping, and those which use vacancy diffusion (Yaldram & Binder 1991).

The regular solution model of an alloy is equivalent to a mean field approximation (MFA) for the binding of an atom in the solid, in which the enthalpy of a random mixture of N_{a} type A atoms and N_{b} type B atoms ($N = N_{\mathrm{a}} + N_{\mathrm{b}}$ total atoms) is given by

$$H = \frac{N_{\mathrm{a}}^2}{2N} mE_{\mathrm{aa}} + \frac{N_{\mathrm{b}}^2}{2N} mE_{\mathrm{bb}} + \frac{N_{\mathrm{a}} N_{\mathrm{b}}}{N} mE_{\mathrm{ab}} = \tfrac{1}{2}Nm\,[x^2 E_{\mathrm{bb}} + (1-x)^2 E_{\mathrm{aa}} + 2x(1-x)\,E_{\mathrm{ab}}],$$

where m is the coordination number and x is the atomic fraction of type B atoms. Including the entropy of the mixture

$$S = k \ln [N_{\mathrm{a}}!\,N_{\mathrm{b}}!/N!] \approx -kN[x \ln(x) + (1-x)\ln(1-x)]$$

gives the Gibbs free energy

$$G = N\{-\tfrac{1}{2}m[x^2(E_{\mathrm{aa}} + E_{\mathrm{bb}} - 2E_{\mathrm{ab}}) + 2x(E_{\mathrm{ab}} - E_{\mathrm{aa}}) + E_{\mathrm{aa}}] \\ + kT[x \ln(x) + (1-x)\ln(1-x)]\}$$

or, for the symmetric case $E_{\mathrm{aa}} = E_{\mathrm{bb}}$,

$$G = N\{-\tfrac{1}{2}m[\alpha(x^2 - x) + E_{\mathrm{aa}}] + kT[x \ln(x) + (1-x)\ln(1-x)]\}.$$

The form of the Gibbs energy at low temperatures is shown in figure 2, showing two minima either side of a central maximum. It is the position of the minima that

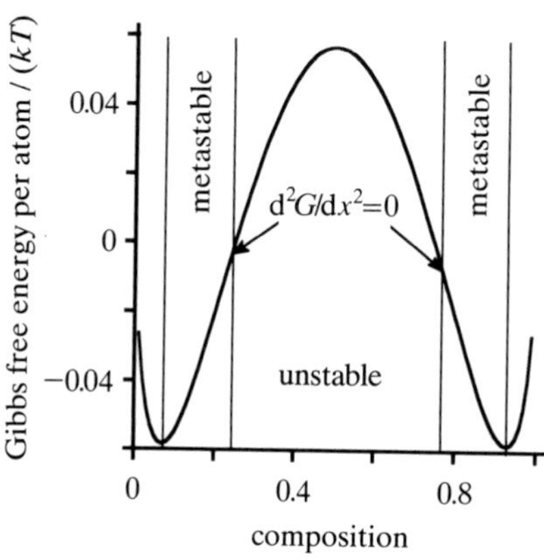

Figure 2. Gibbs free energy curve for a regular solid solution, under conditions where phase separation may occur. The region between the minima in the curve gives the miscibility gap. Within the $\mathrm{d}^2G/\mathrm{d}x^2 = 0$ points on the curve, the solution is unstable and can decompose through a spinodal reaction. Outside these points, the solution is metastable.

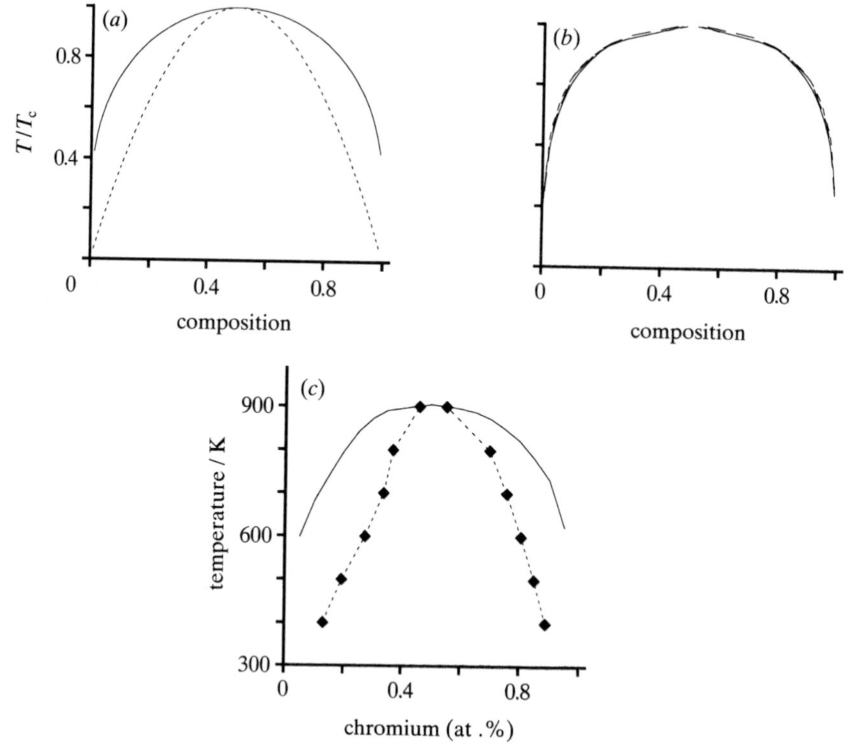

Figure 3. Phase diagrams for (*a*) the regular solution model, (*b*) the Ising model (----, simple cubic; ——, body centred cubic; ——, face centred cubic) and (*c*) the iron–chromium system, as calculated using the ThermoCalc thermodynamical database. In (*a*) and (*c*) the dotted line represents the spinodal region.

defines the solubility limits for the alloy, since between these compositions, the Gibbs energy can be decreased by phase separation. By finding the minima of G, the equation for the solubility limits is found to be

$$T = m\,\alpha(2x-1)/2k \ln[x/(1-x)].$$

Table 1. *Critical energy parameters (α/kT_c) expected from the mean field model, compared with those calculated from the simulation*

lattice	simple cubic	body-centred cubic	face-centred cubic
coordination number (m)	6	8	12
expected critical parameter	0.75	0.5	0.33
calculated critical parameter	0.86	0.62	0.40

The critical temperature, which is the top of the miscibility gap, can be calculated from this equation by finding the limit as the composition tends to $x = 0.5$

$$T_c = \lim_{x\to\frac{1}{2}} m\frac{\alpha(2x-1)}{2k\ln[x/(1-x)]} = \frac{m\alpha}{4k}.$$

The miscibility gap can now be re-expressed in terms of the critical temperature (figure 3*a*)

$$T/T_c = (4x-2)/\ln[x/(1-x)].$$

The solubility limits for the Ising model are found by swapping atoms between two lattices of equal numbers of atoms, both initially set to 50% composition. For low values of the energy parameter α/kT, equivalent to a high model temperature, the two lattices maintain the same composition. At lower temperatures (larger α/kT values) the compositions of the two lattices diverge, reaching equilibrium compositions which are equivalent to the minima in the Gibbs energy. The phase diagrams produced in this way for both a simple cubic lattice and a face-centred cubic lattice are shown in figure 3*b*, where temperatures are plotted as a fraction of the critical temperature for each model. Both models produce essentially the same miscibility gap, which is similar in shape to that expected from the mean field theory. There were some discrepancies, however, between the expected value of the critical energy parameter, and that calculated from the simulation, as shown in table 1. The differences are probably due to the nature of the mean field approximation in describing the nearest neighbour atomic bonding. In the simulations of phase separation, the critical energy parameter was taken to be that calculated from the Ising model, rather than the value expected from the mean field theory.

4. Spinodal decomposition

Alloys with compositions in the centre of the miscibility gap shown in figure 3 are unstable, that is to say that small random fluctuations in local composition will reduce the free energy, and so these variations will be amplified. Under these conditions, phase separation occurs by spinodal decomposition, and the region of instability, delimited by the points on the Gibbs free energy curve for which $d^2G/dx^2 = 0$, is termed the chemical spinodal. Additional energy terms, due to strain for example, are ignored in this treatment. The equation for the chemical spinodal in the mean field model is given by

$$T/T_c = 4x(1-x)$$

and this is shown in figure 3*a*. The phase diagram calculated for the iron–chromium system using the ThermoCalc thermodynamic database (figure 3*c*) shows a broad miscibility gap and a well-defined spinodal region (critical temperature $T_c = 900$ K). Iron–chromium is an ideal model system for the study of spinodal decomposition, and is also important within the context of engineering materials, for example in the

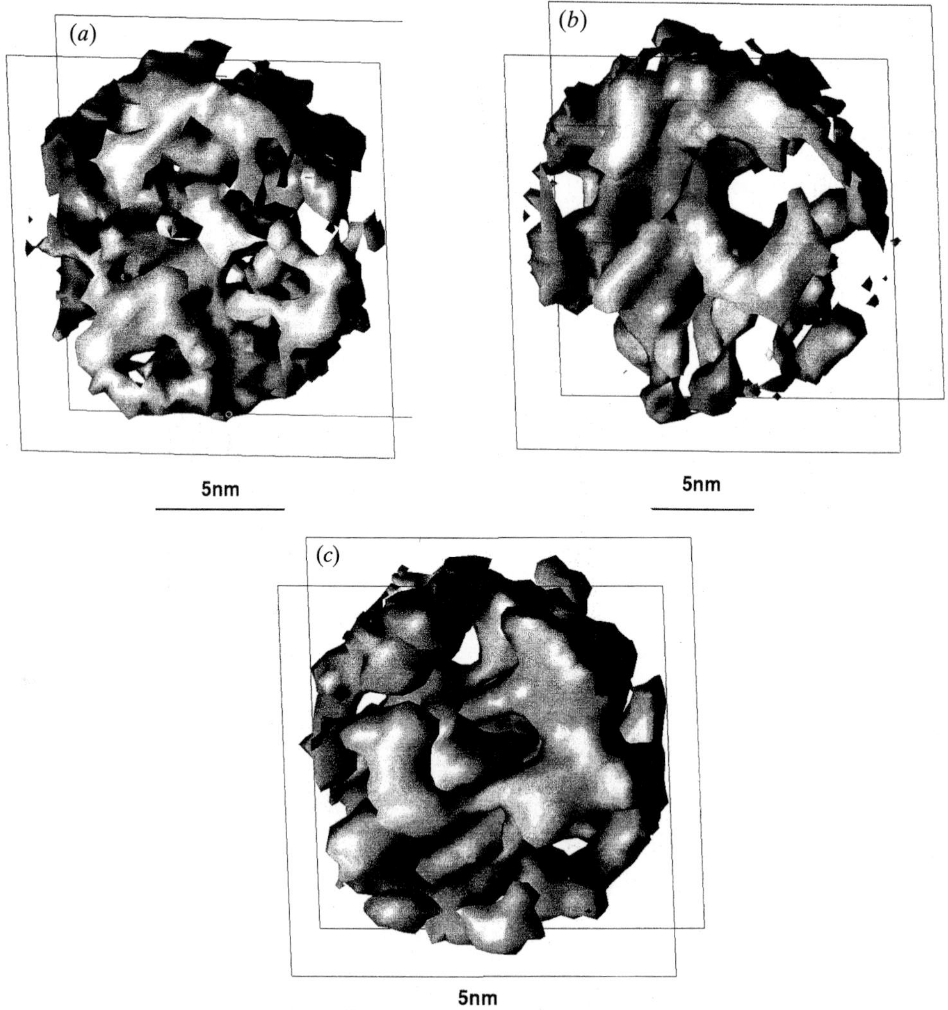

Figure 4. Isosurface reconstructions from the POSAP analysis of Fe–Cr(45 at. %) aged at 500 °C for (a) 24 h, (b) 100 h and (c) 500 h, showing the development in scale during ageing. Note the slight difference in scale in (c).

nuclear and chemical industries. The phases formed during decomposition have a small lattice mismatch (0.6 %), leading to minimal coherency strain. Spinodal decomposition in this system is largely free from competing reactions: the only other phase present (σ) forms too slowly to be observed under most conditions. The present work forms part of an extensive study of the phase separation in iron–chromium by atom probe and POSAP analysis, comparing the experimental results with the Monte Carlo simulations (Hyde *et al.* 1992). A series of isosurface reconstructions from Fe–Cr(45 at. %) aged for a range of times at 773 K is shown in figure 4, and indicates how the scale of the microstructure present in this alloy develops in the early stages of phase separation. Since the miscibility gap in the iron–chromium system is not quite symmetrical, the Fe–Cr(45 at. %) alloy produces a 50 % volume fraction of the chromium rich phase. At these high volume fractions, the resulting microstructure has a complex, interconnected morphology.

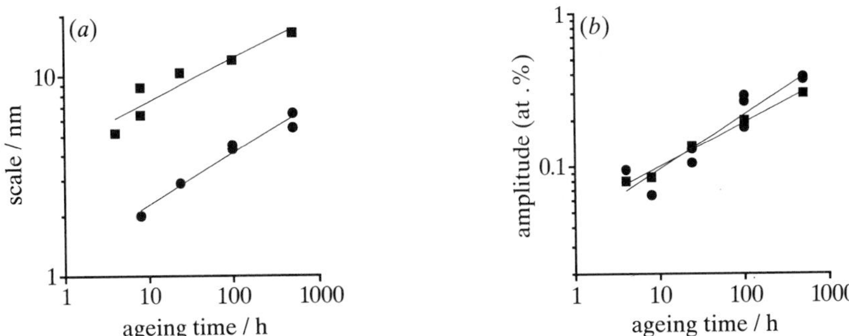

Figure 5. Kinetics of phase separation in Fe–Cr(45 at. %) aged at 500 °C as measured by conventional atom probe and position-sensitive atom probe techniques: (*a*) growth of wavelength of the microstructure; (*b*) development of composition amplitude. ●, POSAP; ■, AP.

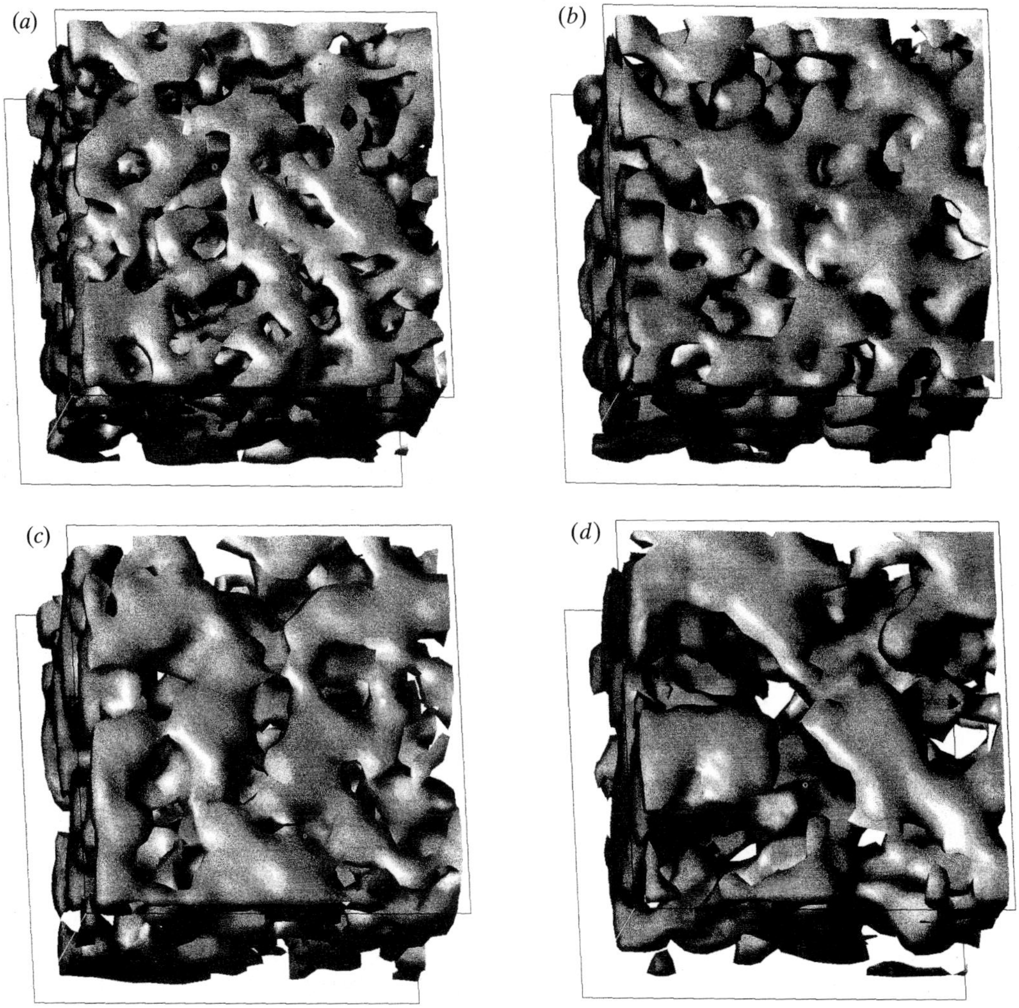

(*a*) (*b*)

(*c*) (*d*)

Figure 6. Isosurface representations of the microstructures in the Monte Carlo simulation of phase separation after (*a*) 10 Monte Carlo steps (MCS), (*b*) 100 MCS, (*c*) 1000 MCS, and (*d*) 10000 MCS. One MCS represents a complete sweep through the lattice, although the random selection means that given sites may be missed in any given step.

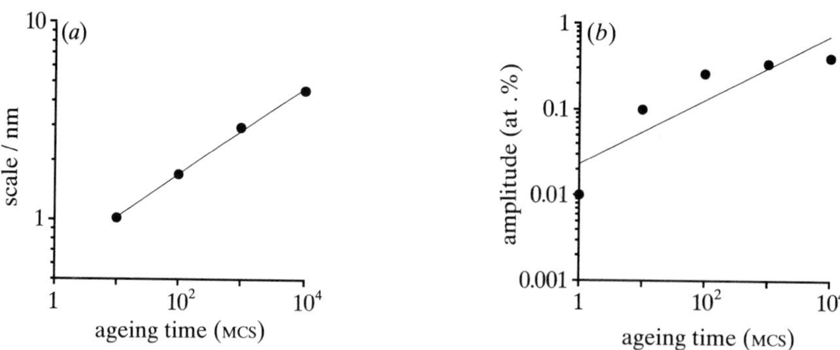

Figure 7. Kinetics of phase separation in the Monte Carlo simulation: (*a*) change in wavelength; (*b*) development of composition amplitude.

The conventional formulation of the theory of spinodal decomposition is due to Cahn & Hilliard (1958). In their model, they consider the local composition at given points in the alloy as a continuous variable, and the composition fluctuations as a series of Fourier components. The process of decomposition is treated as the selective growth of these components of composition fluctuation, and this theory predicts that in the early stages the composition amplitude grows exponentially, with a small range of wavelengths dominating. In the earliest stages, therefore, the wavelength of the microstructure should remain approximately constant. The experimentally observed growth behaviour is usually different from that predicted by this model, as shown for the case of Fe–Cr(45 at. %) in figure 5. Even from the earliest stages of phase separation accessible to the atom probe techniques, the wavelength is observed to grow (figure 5*a*). The observed growth of the structure can be fitted to a t^a power law, with a time scaling of $a = 0.21$ for the atom probe data and $a = 0.26$ for the POSAP results. There is no indication of an exponential increase in the composition amplitude (figure 5*b*), even though it was possible to determine the magnitude of the fluctuations down to below 10 at. % in structures with wavelengths of approximately 1 nm. Fitting a t^a power law once again, the measured time exponent from the atom probe is $a = 0.29$, and the POSAP results give $a = 0.36$.

An equivalent set of measurements from the Monte Carlo simulation of phase separation in the Ising model is shown in figures 6 and 7. Modelling was carried out using a 50% atomic fraction, which yields a 50% volume fraction of the two phases after separation, equivalent to the case of the Fe–Cr(45 at. %) alloy analysed experimentally. A modelled temperature of 800 K was established by comparing the critical energy parameter of the model with the critical temperature of the iron–chromium system calculated from the ThermoCalc database (900 K). The development of the microstructure on ageing can be seen clearly in the sequence of isosurface reconstructions shown in figure 6. By analysing the atomic positions from the model in exactly the same way as for the POSAP data, the rate of growth of the wavelength and the amplitude of composition fluctuations can be determined from the simulation (figure 7*a*, *b*). These were fitted to a t^a power law as for the experimental data, and it was found that the increase in scale followed a $t^{0.22}$ dependence. The composition fluctuations grew in amplitude with a $t^{0.36}$ dependence, although the fit in this case is not as good. In both cases the scaling behaviour is in good agreement with the experimental observations. The results shown here were calculated using a simple cubic lattice but simulations based on a more realistic body-centred cubic lattice showed few quantitative differences. Since spatial information present in the

POSAP data allows a full three-dimensional reconstruction of the atomic-scale chemistry present in the material, it is also possible to perform topological and morphological analysis of the microstructure, as discussed by Cerezo *et al.* (1991). This permits the ageing process to be characterized still further. For example, the fractal dimension of the interface between chromium-rich and iron-rich phases can be measured, and the change of dimension with ageing time explored. Hyde *et al.* used these techniques for a detailed comparison between the ageing of a number of iron–chromium alloys and the results from the Monte Carlo simulation. These authors found that the simulations were able to represent accurately the development of microstructural features.

5. Nucleation and growth

Outside of the spinodal region of the phase diagram (figure 3*a*), but still within the miscibility gap, the solid solution is metastable and may undergo phase separation by nucleation and growth. Composition fluctuations which are small in amplitude and extent are unstable and tend to decay, but once a nucleus of the second phase reaches a critical size, it is stable and can continue to grow. In the conventional model for this process, the nucleus is treated as a perfect sphere of radius r. Obviously, the nucleus cannot be a perfect sphere at this scale, and this is simply a continuum approximation of the problem. The free energy of the nucleus is considered to consist of a favourable volume term, due to the bonding of like atoms in the nucleus, and an unfavourable surface term which is sometimes called surface tension but may be viewed principally in terms of the binding between unlike atoms across the interface. Since the volume term varies as r^3 and the surface term as r^2, a potential barrier exists for the growth of the nucleus. When the top of the barrier is reached (at the critical size) it is energetically favourable for the nucleus to continue to grow. In practise the distinction between the regimes of spinodal decomposition and nucleation is blurred, since close to the spinodal limit, both the energy barrier for nucleation and the critical nucleus size will be small.

Once again, there exists a good model alloy for the study of nucleation and growth: copper–cobalt. The thermodynamics of the system are well established, and the phase diagram has a broad miscibility gap and no intermediate phases. Solid solubilities at both sides of the phase diagram are small so precipitates in these alloys should be almost pure cobalt or copper. These have the same face-centred cubic crystal structure, and the lattice mismatch is small (below 2%). A sequence of POSAP analyses of Cu–Co(1.0 at. %) aged at 723 K is shown in figure 8. In the earliest stages (10 min) the distribution of cobalt atoms appears to be random. After ageing for $\frac{1}{2}$–1 h, clusters of cobalt atoms (embryos) can be seen, and statistical analysis shows the distribution to be non-random. Ageing for 24 h produces a distribution of nuclei approximately 1.5 nm in diameter of almost pure cobalt (over 90 at. %).

Is it possible for a simple Monte Carlo simulation to adequately represent this behaviour? At first, it might seem not, since whatever the size of a nucleus, moving a cobalt atom from the matrix to the surface of the nucleus will always decrease the free energy if the mean composition is in the miscibility gap. However, the formation of a nucleus greatly reduces the entropy of the system, and although this is not considered explicitly in the simulation, the stochastic nature of the algorithm effectively incorporates entropy in the model. The behaviour of the Ising model can therefore be treated using standard gibbsian thermodynamics. Consider a system of

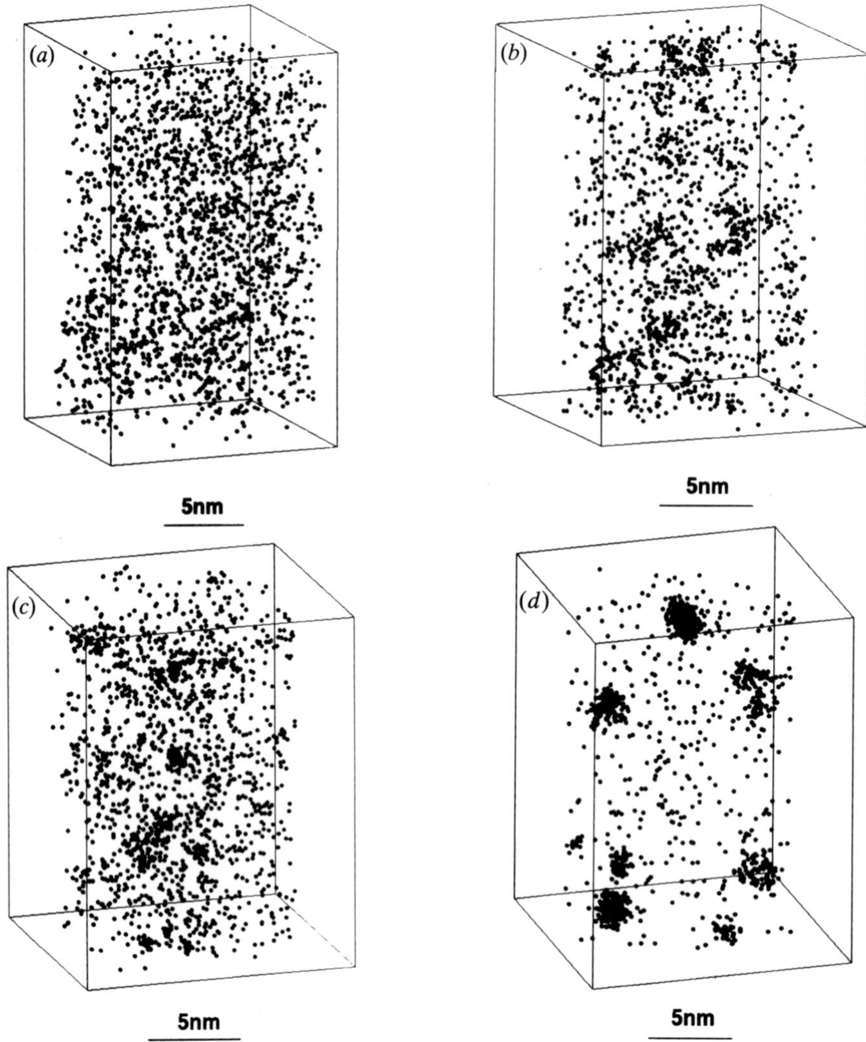

Figure 8. Distribution of cobalt atoms from the POSAP analysis of Cu–Co(1 at. %) aged at 723 K for (*a*) 10 min, (*b*) 30 min, (*c*) 2 h and (*d*) 24 h, showing the growth of cobalt precipitates.

N atoms, as before, in which a nucleus of N_c atoms of composition x_c is formed. The composition x_m of the remaining N_m matrix atoms is given by

$$N_m(x - x_m) = N_c(x_c - x).$$

The change in free energy in forming the nucleus is therefore

$$\Delta G = N_c[g(x_c) - g(x)] + N_m[g(x_m) - g(x)] + \tfrac{1}{2}N_i\, m[E_{bb} - E_{ab}]\, x_c,$$

where $g(x)$ is the Gibbs free energy per atom of a regular solution of composition x, and N_i is the number of B atoms which are at the interface of the cluster. It is assumed that on average the interface atoms have half of their bonds to the precipitate, and half to the matrix. Using the approximation of a spherical particle, the number of interface atoms can be calculated as

$$N_i = \sqrt[3]{(36\pi N_c^2)}.$$

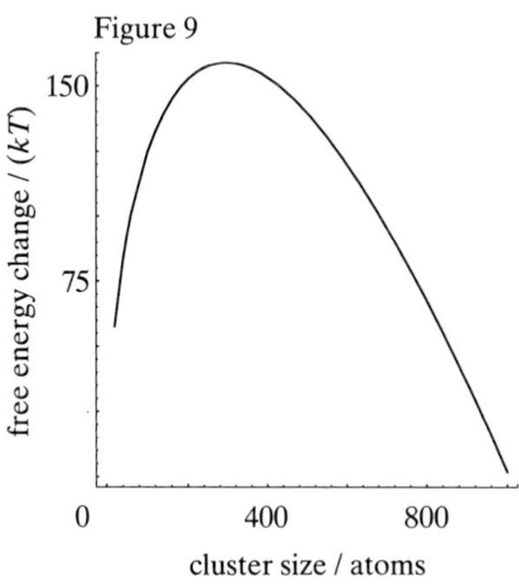

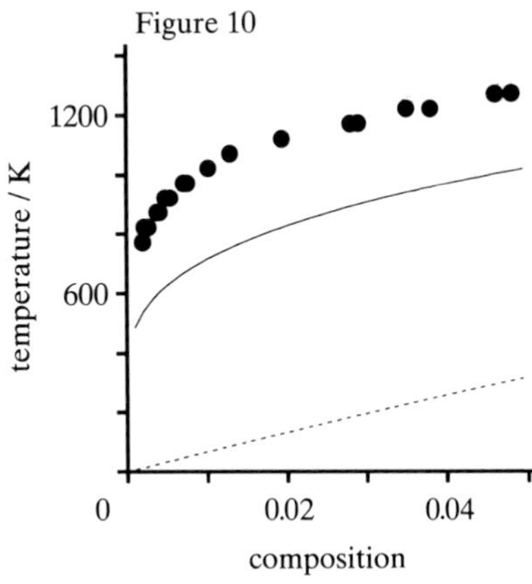

Figure 9. Variation in free energy gain between a random solid solution and a distribution of clusters as a function of the cluster size for a temperature of $0.43\,T_c$ and a composition of 4 at.%. The energy barrier to the growth of nuclei is clearly seen, and under these conditions the critical nucleus size is just over 300 atoms.

Figure 10. Copper–cobalt phase diagram from the data compiled by Nishizawa & Ishida (1984) (●) compared with the miscibility gap (——) and chemical spinodal (····) for the Ising model, assuming a critical temperature of $T_c = 1680$ K.

In the limit of $N_m \gg N_c$, the change in the free energy of the matrix is given by

$$N_m[g(x_m)-g(x)] = N_m[x-x_m]\left[\frac{dg}{dx}\right]_{x=x_m} = N_c[x_c-x]\left[\frac{dg}{dx}\right]_{x=x_m}$$

Substituting for the critical temperature from above gives an expression for the total free energy change:

$$\Delta G = N_c\left[g(x_c)-g(x)+[x_c-x]\left[\frac{dg}{dx}\right]_{x=x_m}\right]+x_c\sqrt[3]{(36\pi N_c^2)}\,k\,T_c.$$

The variation in free energy as a function of cluster size is shown in figure 9 for a temperature of $0.43\,T_c$ and a composition of 4 at.%, calculated assuming that the nuclei formed are of the composition which gives a minimum in the Gibbs free entry ($x \approx 1-e^{-2T_c/T}$). There is an energy barrier of approximately 170 kT to the growth of nuclei, and the critical nucleus size is below 300 atoms.

Since the copper–cobalt phase diagram does not permit the assignment of a critical temperature due to an intervening peritectic reaction, a critical temperature of 1680 K was used for the Monte Carlo simulations, based on the estimate by Chisholm (1986) who fitted a sub-regular solution model to the alloy solubility data for temperatures in the range 800–1400 K. However, the Ising model is essentially a regular solution, and the use of the critical temperature from a sub-regular model leads to a discrepancy between the phase diagrams of model and the alloy. A comparison of the miscibility gap in the Ising model, assuming a critical temperature of $T_c = 1680$ K, and some of the solubility data for copper–cobalt alloys compiled by Nishizawa and Ishida (1984) is shown in figure 10. The degree of undercooling for a given composition will be underestimated in the simulation, and the undercooling for a 1% alloy is approximately equal to that of the simulation at a 4% atomic fraction.

The results of a Monte Carlo simulation for a 4% atomic fraction at a temperature of $0.43\,T_c$ (equivalent to 723 K for $T_c = 1680$ K) are shown in figure 11. A graph of

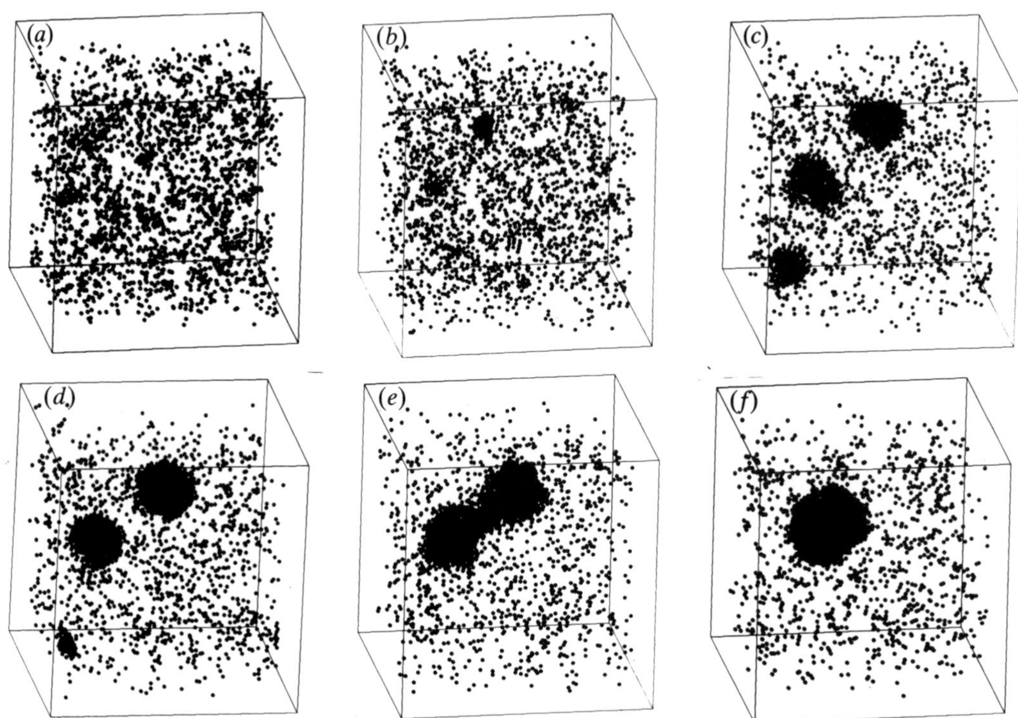

Figure 11. Atom positions from a Monte Carlo simulation of ageing in an Ising model, for an atomic fraction of 4% and a temperature of 0.43 T_c after (*a*) 2000 MCS, (*b*) 2750 MCS, (*c*) 5750 MCS, (*d*) 8500 MCS, (*e*) 14500 MCS and (*f*) 20000 MCS.

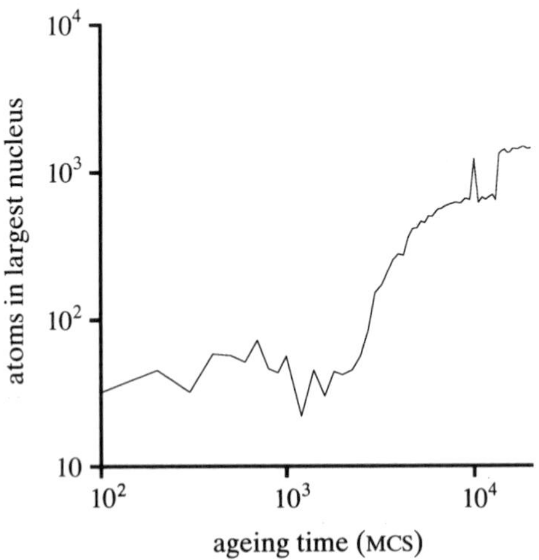

Figure 12. Graph showing the size of the largest cluster as a function of ageing time in the simulation shown in figure 10.

the size of the largest cluster as a function of ageing time in the simulation is given in figure 12. From the initial random arrangement of solute atoms, embryos quickly form (figure 10*a*). There is then a period in which none of the embryos within the modelled volume exceeds the critical size. At this point the phase separation appears static, although if it were possible to model a much larger volume of material, a slow increase in the density of critical nuclei should be observed. Finally a nucleus is formed within the model which exceeds the critical size and this grows rapidly (figure 10*b*). Two more nuclei are formed soon after (figure 10*c*) but one of these redissolves almost immediately as the other two nuclei grow (figure 10*d*). At later stages the two remaining nuclei coalesce (figure 10*e*) to form a single cobalt-rich particle (figure 10*f*). Measurements of the composition of the nuclei from the simulation show them to be essentially 100% cobalt as soon as they are formed. The phase decomposition in the model is therefore a classical one, contrary to the conclusion drawn by Chisholm & Laughlin (1987) from simulations carried out with similar parameters. That is not to say, however, that the transformation occurs classically right up to the spinodal. It is intended to perform modelling across a range of temperatures and compositions to observe how the transformation behaviour changes progressively from classical nucleation to spinodal decomposition.

It is clear from this example of a simulated ageing, that the simple model is able to reproduce many of the features which we expect to see in the process of homogeneous nucleation and growth, but at the atomic level. The 'incubation period', nucleation barrier and Ostwald ripening all seem to be represented in the particular example shown here.

6. Conclusions

Despite the extreme simplicity of the Monte Carlo simulation described here, it appears to model quite successfully many of the features of the early stages of phase transformations which are observed experimentally using the atom probe and POSAP, covering the timescales which are typical of heat treatments or thermal ageing of alloys. In modelling of spinodal decomposition, the simulation has provided quantitative fits to the dynamics of phase separation in iron–chromium alloys, including the development of microstructural features. Initial results on the modelling of nucleation and growth appear promising, although it may be some time before the same quantitative agreement is attained. Given the success of such a basic simulation method, it is hoped that further development, for example by the use more realistic interatomic potentials, will extend the usefulness of the technique to a wider range of materials, and allow the accurate modelling of the early stages of other phase transformations, such as ordering and the conditional spinodal reaction.

A.C. is grateful to The Royal Society and Wolfson College, Oxford for support during the course of this work. J.M.H. and R.P.S. thank the Science and Engineering Research Council for financial support in the form of studentships. The studies of iron–chromium are sponsored in part by the Division of Materials Sciences, U.S. Department of Energy, under contract DE-AC05-84OR21400 with Martin Marietta Energy Systems, Inc. The authors are grateful to Dr M. F. Chisholm, Oak Ridge National Laboratory for the supply of the copper–cobalt alloys used in this work. This research is being funded by the Science and Engineering Research Council under grant no. GR/H/384845.

References

Cahn, J. W. & Hilliard, J. E. 1958 Free energy of a non-uniform system I. Interfacial free energy. *J. chem. Phys.* **28**, 258–267.

Cerezo, A., Godfrey, T. J. & Grovenor, C. R. M. 1989 Materials analysis with a position-sensitive atom probe *J. Microsc.* **154**, 215–225.

Cerezo, A., Hetherington, M. G., Hyde, J. M. & Miller, M. K. 1991 A topological approach to materials characterisation. *Scripta metall.* **25**, 1435–1440.

Chisholm, M. F. & Laughlin, D. E. 1987 Decomposition in alloys: an overview. In *Phase transformations 87* (ed. G. W. Lorimer), p. 17. London: Institute of Metals.

Hyde, J. M., Cerezo, A., Hetherington, M. G., Miller, M. K. & Smith, G. D. W. 1992 Three-dimensional characterisation of spinodally decomposed iron–chromium alloys. *Surf. Sci.* **266**, 370–377.

Metropolis, N., Rosenbluth, A. W., Teller, A. H. & Teller, E. 1953 Equation of state calculations by fast computing machines. *J. chem. Phys.* **21**, 1087–1092.

Miller, M. K. & Smith, G. D. W. 1989 *Atom probe microanalysis: principles and applications to materials science.* Pittsburgh, U.S.A.: Materials Research Society.

Müller, E. W., Panitz, J. A. & McLane, S. B. 1968 The atom-probe field ion microscope *Rev. Sci. Instrum.* **39**, 83–86.

Nishizawa, T. & Ishida, K. 1984 The Co–Cu (cobalt–copper) system. *Bull. Alloy Phase Diagrams* **5**, 161–165.

Yaldram, K. & Binder, K. 1991 Spinodal decomposition of a two-dimensional model alloy with mobile vacancies. *Acta. metall.* **39**, 707–717.

Intermolecular interactions and the nature of orientational ordering in the solid fullerenes C_{60} and C_{70}

By Ailan Cheng[1], Michael Klein[1], Michele Parrinello[2] and Michiel Sprik[2]

[1] *Department of Chemistry and Laboratory for Research on the Structure of Matter, University of Pennsylvania, Philadelphia, Pennsylvania 19104–6323, U.S.A.*

[2] *IBM Research Division, Zurich Research Laboratory, 3303 Rüschlikon, Switzerland*

We have proposed an intermolecular potential for C_{60} molecules that not only reproduces the correct low-temperature structure, but also correlates a wide range of experimental properties, including the molecular reorientational time in the room-temperature rotator phase, the volume change at the orientational ordering transition, and the librational frequencies in the low-temperature phase. The low-pressure phases in solid C_{70} have been explored using constant-pressure molecular dynamics and an intermolecular potential derived from one that gives an excellent account of the properties of solid C_{60}. The molecular dynamics calculations predict three low-pressure phases: a high-temperature rotator phase, a partly ordered phase with trigonal symmetry, and an ordered monoclinic phase. The calculations on C_{70} were carried out on a cluster of IBM RS/6000s, operating in parallel.

1. Introduction

Our understanding of solid fullerenes is largely derived from detailed experimental studies on Buckminsterfullerenes (Taylor *et al.* 1990; David *et al.* 1991; Heiney *et al.* 1991; Johnson *et al.* 1991; Neumann *et al.* 1991; Sachidanandum & Harris 1991; Tycko *et al.* 1991*a*,*b*; Yannoni *et al.* 1991; Copley *et al.* 1992; David *et al.* 1992; Heiney *et al.* 1992; van Loosdrecht *et al.* 1992; Shi *et al.* 1992). This new allotrope of carbon, composed of C_{60} clusters, behaves as a typical molecular crystal. The room-temperature structure is a face-centred cubic (FCC) rotator phase, which transforms to an orientationally ordered state at 250 K. The low-temperature structure is simple cubic, with space group $Pa\bar{3}$ (David *et al.* 1991). The observed $Pa\bar{3}$ structure has electron-rich short hexagon bonds facing electron-poor pentagon centres of adjacent molecules. Static structure optimization and molecular dynamics calculations based on atom–atom potentials (Guo *et al.* 1991) have failed to explain the experimental low-temperature $Pa\bar{3}$ structure. Instead, they predict a tetragonal, rather than cubic, low-temperature phase. The discrepancy in the ordered phase is in contrast to the surprisingly accurate description provided by the atom–atom potential of the structure and dynamics of the disordered high-temperature phase (Cheng & Klein 1991). Solid C_{70} has received less attention. Because of the elongated 'rugby ball' shape (Fowler *et al.* 1991; Andreoni *et al.* 1992), the phase diagram of C_{70} will likely be more complex. Indeed, X-ray diffraction measurements (Vaughan *et al.* 1992)

Phil. Trans. R. Soc. Lond. A (1992) **341**, 327–336

Printed in Great Britain

© 1992 The Royal Society

indicate a FCC high-temperature phase, plus phase transitions at about 340 K and 280 K. Neither the molecular ordering nor the lattice symmetry of the low-temperature phases have yet been resolved.

In this paper, we review previous attempts to construct an empirical intermolecular potential for C_{60} molecules. Potentials now exist that not only reproduce the correct low-temperature structure of C_{60}, but also correlate a range of experimental data, including the molecular reorientational time in the room-temperature rotator phase (Johnson *et al.* 1991; Tycko *et al.* 1991), the volume change at the orientational ordering transition (Heiney *et al.* 1992), and the librational frequencies in the low-temperature phase (Copley *et al.* 1992). The potential model has been extended to treat C_{70} and the nature of the low-pressure phases of the solid have been explored using constant-pressure molecular dynamics (MD) calculations (Nosé & Klein 1983; Parrinello & Rahman 1980). The number of interaction sites on a C_{70} molecule is *ca.* 100, so that each pair potential requires evaluation of 100×100 terms. This type of intensive MD calculation is eminently suited to parallel processing. The results of the simulations, carried out on a cluster of IBM RS/6000 workstations, yield predictions concerning the structure of solid C_{70} in its room-temperature phase and the ground state.

2. An intermolecular potential for solid C_{60}

To achieve the level of accuracy mentioned above for C_{60}, it was necessary to explicitly include interaction sites on the electron-rich short C–C bonds, plus an electrostatic contribution to the intermolecular potential. The distinction between the two bond types can be justified qualitatively by experimental (David *et al.* 1991) and theoretical evidence (Fowler *et al.* 1990, 1991) indicating that the conjugation of the π bond network is incomplete; the length of a bond shared by two hexagons is shorter than bonds of the pentagons. The explicit representation of multiple bonds in the modelling of intermolecular interactions has an important precedent in the study of solid nitrogen, where it proved necessary to place a third interaction centre in the middle of the triple bond (Raich & Mills 1971).

The situation for C_{60} at low pressure is akin to high-pressure nitrogen, because carbon atoms at the sites where the C_{60} molecules make contact are pressed together by the large cohesive forces of all the other atoms. Thus, local C–C contacts in solid C_{60} are actually much closer than those between lamella in graphite. In C_{60}, contacts are optimized when a short C–C bond is wedged into the centre of a pentagon and the distance between bonds maximized (David *et al.* 1991). On the other hand, the spurious tetragonal structure favoured by the atom–atom potential model is characterized by crossed short bonds at minimum separation (Guo *et al.* 1991; Cheng & Klein 1992*a*).

In the interacting bond model of Sprik *et al.* (1992), the 60 Lennard–Jones carbon sites of C_{60} were supplemented with 30 sites D, located at the centres of the double bonds. The molecule was treated as a rigid framework of atoms and bonds, with the geometry taken from Fowler *et al.* (1991) and Feuston *et al.* (1991). In addition, the modelling of solid nitrogen suggested a way to further refine the potential. The nitrogen molecule, as a consequence of the triple bond, carries an appreciable electrostatic quadrupole moment. The corresponding intermolecular coupling in C_{60}, which involves $l = 6$, 10, 18, ... spherical harmonics, cannot be ignored in a description of the solid (Michel *et al.* 1992). To account for this important electrostatic component, Sprik *et al.* (1992) assigned a negative bond charge q_D to

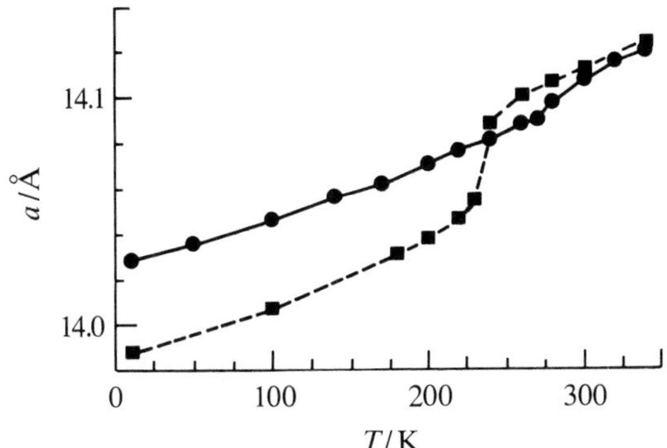

Figure 1. The squares and dots indicate the calculated temperature dependence of the cubic lattice parameters of solid C_{60} for the models of Sprik *et al.* (1992*a*, *b*) and Lu *et al.* (1992) respectively.

the D sites, and a compensating positive charge $q_C = -\frac{1}{2}q_D$ to the C atoms. As a result, the pentagons acquired a total charge of $5q_C$. The D sites are attracted by this net accumulation of positive charge, which, in turn, enhances the stability of the observed $Pa\bar{3}$ structure. In this formulation of the intermolecular potential, the electrostatic contribution to the orientational ordering is important but secondary to the short-range repulsion on the D sites. The potential parameters $\sigma_{CD} = 3.4$ Å†, and $\sigma_{DD} = 3.6$ Å were fixed and ϵ and q_D adjusted to yield reasonable values for the transition temperature (Dworkin *et al.* 1991), the cohesive energy (Pan *et al.* 1991), the reorientational relaxation time in the disordered phase (Johnson *et al.* 1991; Tycko *et al.* 1991*a*, *b*), plus the librational spectrum in the ordered phase (Copley *et al.* 1992). With a value of $q_D = -0.35e$ and $\epsilon = 15$ K, there is reasonable overall agreement among all four properties.

The calculations required for the parameter fitting were performed using a constant pressure-constant temperature molecular dynamics algorithm (Nosé & Klein 1983; Parrinello & Rahman 1980), and systems of 32 or 108 molecules, replicated by periodic boundary conditions (Allen & Tildesley 1987). The length of a run at a single state point was typically between 50 and 100 ps. The calculated structural transformation is an abrupt event (see figure 1) occurring around 225 K, which is somewhat lower than the experimental value of *ca.* 250 K. The transformation is clearly first order, with a modest hysteresis (Sprik *et al.* 1992). The calculated discontinuity agrees well with recent measurements (Heiney *et al.* 1992). The total binding energy of the model compares favourably with the measured enthalpy of formation $\Delta H = 171$ kJ mol^{-1} (Pan *et al.* 1991). Figure 2 contains snapshots of typical configurations from the high and low-temperature phases of C_{60}.

The effect of pressure on the orientational transition temperature T_t was also investigated with the result $(dT_t/dP) = 12 \pm 4$ K kbar^{-1}‡, which is in reasonable accord with the experimental coefficient 11.7 K bar^{-1} (Kriza *et al.* 1991). The spectral density of the librational modes in the ordered phase at 100 K was calculated. The band of librational frequency was recently measured experimentally to be centred around 20 cm^{-1} (Copley *et al.* 1992), in fair agreement with the present model, which yields 15 cm^{-1}. The relaxation time of orientational correlations in the disordered phase obtained from NMR studies at 300 K was determined by Tycko *et al.* (1991*a*, *b*) and Johnson *et al.* (1992) as 12 ps and 8 ps respectively. The result from the MD calculations was 9 ± 2 ps.

† Å = 10^{-10} m = 10^{-1} nm. ‡ 1 bar = 10^5 Pa.

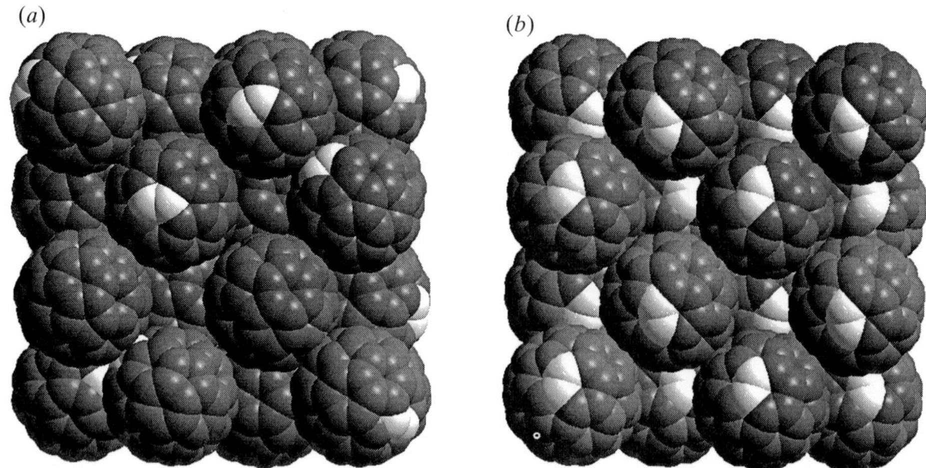

Figure 2. Instantaneous configurations taken from molecular dynamics simulations of the fullerene C_{60}. (*a*) The room-temperature rotator phase. (*b*) The low-temperature ordered phase.

The performance of the intermolecular potential model for C_{60} discussed above suggests the need not only for explicit representation of interaction sites on the short electron-rich bonds, but also electrostatic interactions. This completely empirical intermolecular potential is able to reproduce most experimental quantities rather well. The discrepancies that remain indicate a need for increased anisotropy of the molecules; a deficiency that could easily be corrected by further refining the parameters. Unfortunately, at the present time, an *ab initio* approach to solid C_{60} is difficult because of the large number (240) of carbon atoms in the $Pa\bar{3}$ unit cell (Feuston *et al.* 1991).

3. An alternative potential model for C_{60}

The structure of the orientationally ordered low-temperature phase of solid C_{60}, as discussed above, cannot be explained solely in terms of Lennard–Jones interactions between carbon atoms (Cheng & Klein 1992*a*, *c*). Recently, Lu *et al.* (1992) proposed an alternative extension of the atom–atom potential, in which positive and negative charge sites were positioned at the centres of the long and short carbon–carbon bonds respectively. In mean field theory, this version of the bond charge model exhibits a phase transition from a rotator phase to the experimentally observed ordered $Pa\bar{3}$ structure at about the correct temperature. Thus, the electrostatic contribution from bond charges alone proved sufficient to stabilize the $Pa\bar{3}$ structure.

We have examined the properties of the potential model proposed by Lu *et al.* (1992), using constant-pressure molecular dynamics (Nosé & Klein 1983). Figure 1 compares the cubic lattice parameters predicted for the models of Sprik *et al.* (1992) and Lu *et al.* (1992). The rotational ordering transition in the latter model occurs a little above the experimentally observed temperature, but is accompanied by only a marginal volume contraction. By contrast, the discontinuity calculated for the potential of Sprik *et al.* (1992) is much larger and in good agreement with experiment (Heiney *et al.* 1992), although the transition temperature is a little too low. The strikingly different behaviour exhibited by the two models can be rationalized as follows: molecules in the ordered phase of the Sprik *et al.* potential fit together in a lock and key fashion and the unit cell volume is therefore reduced with respect to the rotator phase. This mechanism is far less pronounced for the potential of Lu *et al.* (1992) where the ordering is driven by electrostatic coupling. Apparently, a repulsive contribution arising from the electron density on the short bonds cannot be ignored.

4. Ground state at phase transition of solid C_{70}

As mentioned above, attempts to explain the stability of the $Pa\bar{3}$ ground state of C_{60} by means of simple atom–atom type interactions between C atoms were unsuccessful (Cheng & Klein 1992*a–c*). Extension of the model to include electrostatic interactions involving additional interaction sites on C–C bonds seemed to be required (Sprik *et al.* 1992; Lu *et al.* 1992). We propose herein a generalization of the Sprik *et al.* interaction model for C_{60} to the case of C_{70}. The potential is then used in an exploration of the solid-phase diagram, using constant pressure MD simulation, with no experimental input, other than the assumption of a FCC rotator phase at high temperatures (Vaughan *et al.* 1992).

The initial generalization of the C_{60} potential to C_{70} assumes that all C atoms can still be treated as a single type, with the C_{60} value for σ_C and q_C, even though in C_{70} these sites are no longer strictly equivalent (Fowler *et al.* 1991). Conceptually, the C_{70} molecule can be partitioned in two C_{60}-like 'polar caps', consisting of 20 atoms each, plus an 'equatorial' region of 30 atoms. The polar caps consist of a top pentagon and five hexagons, each sharing a long bond with the top pentagon. The 20 short bonds of these C_{60}-like caps are assigned D sites with the same σ_{DD} as in the C_{60} model and $q = -2q_C$. Of the bonds in the equatorial region, 30 have a length slightly larger than the short D bonds. These intermediate (I) bonds are grouped in five isolated hexagons. The remaining bonds have lengths more similar to the long bonds in C_{60}. On the basis of a length criterion and considerations of charge neutrality, we ignore the longer bonds and only include sites on I bonds with a σ_{II} $(= 3.5 \text{ Å})$ chosen intermediate between the values for C and D sites and a charge $q_I = \frac{1}{2}q_D$. The radii for interactions between dissimilar pairs are determined by the arithmetic mean mixing rule. The value $\epsilon = 15$ K was used for all Lennard–Jones terms. The geometry of the rigid molecular C_{70} frame was taken from Andreoni *et al.* (1992).

According to experiment (Vaughan *et al.* 1992), in the high-temperature rotator phase I the FCC centre-of-mass structure is favoured over hexagonal close packed (HCP). The free energy difference between FCC and HCP is likely to be very small (Guo *et al.* 1991; Cheng & Klein 1992*c*). The simulation was initialized by setting up an orientationally disordered high-temperature state in a cubic MD box, which consisted of $4 \times 4 \times 4$ FCC 4-molecule unit cells with basis vectors $\boldsymbol{a}_{FCC}$, $\boldsymbol{b}_{FCC}$ and $\boldsymbol{c}_{FCC}$. The system size of 256 molecules with $\boldsymbol{L}_a = 4\boldsymbol{a}_{FCC}$ and $\boldsymbol{L}_c = 4\boldsymbol{c}_{FCC}$ was found to be necessary to minimize boundary effects. The simulation system was slowly cooled by applying the constant temperature-constant pressure MD technique (Sprik *et al.* 1992). Energies and forces were determined from a summation over all interaction sites on the 12 nearest and 6 next nearest neighbour molecules. The temperature was decreased in steps of 20 K (occasionally 10 K) after waiting at each state point 40–80 ps. The simulation, which effectively contains more than 30 000 atoms, was executed on a cluster of IBM RS/6000 workstations operating in parallel under the Parallel Virtual Machine (PVM) communication software (public domain) package (G. A. Geist & V. S. Sunderam). With cooling rates mentioned above, the total run time to follow

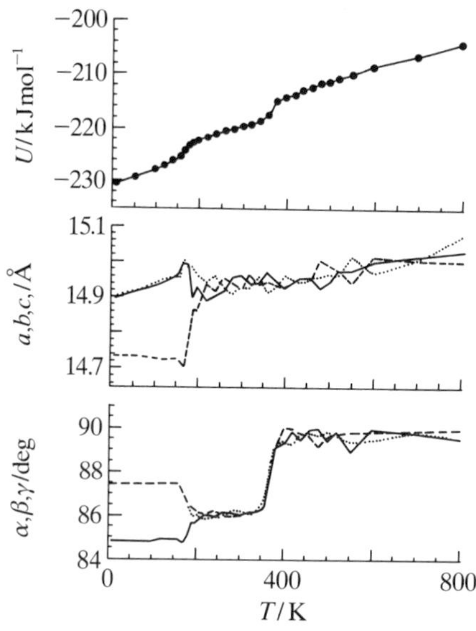

Figure 3. Calculated temperature dependence of selected properties of solid C_{70}. (*a*) The intermolecular potential energy, U; (*b*) MD box lengths plotted as cell vectors $\boldsymbol{a}_{\text{FCC}} = \frac{1}{4}\boldsymbol{L}_{\text{a}}$, $\boldsymbol{b}_{\text{FCC}} = \frac{1}{4}\boldsymbol{L}_{\text{b}}$, $\boldsymbol{c}_{\text{FCC}} = \frac{1}{4}\boldsymbol{L}_{\text{c}}$, and (*c*) angles between cell vectors.

the C_{70} system through two orientational ordering transitions was about three months.

The MD results, shown in figure 3, revealed two phase transitions. The first transition, $T_{\text{I-II}}$, occurred around 380 K, and the second, $T_{\text{II-III}}$, around 200 K. From the temperature dependence of the total interaction energy, U, we estimate the heats of transition as $\Delta H_{\text{I-II}} = 2.8 \pm 0.3$ kJ mol^{-1} and $\Delta H_{\text{II-III}} = 2.5 \pm 0.3$ kJ mol^{-1}. Vaughan *et al.* (1992) observed the I–II transition around 340 K, with $\Delta H_{\text{I-II}} = 2.9 \pm 0.4$ kJ mol^{-1}, and the II–III transition around 280 K, with $\Delta H_{\text{II-III}} = 2.3 \pm 0.4$ kJ mol^{-1}. These results agree surprisingly well with our calculations, the largest discrepancy being the value of $T_{\text{II-III}}$. Vaughan *et al.* quote the FCC lattice constant in phase I at 325 K, just above the I–II phase transition, as $a = 14.96$ Å. At the corresponding point, we find a very similar value (see figure 3).

In constant-pressure molecular dynamics, any breaking of the point group symmetry can be monitored from the shape changes of the MD box (Parrinello & Rahman 1980). Figure 3 also shows the variation in lengths of the cell vectors $\boldsymbol{a}_{\text{FCC}}$, $\boldsymbol{b}_{\text{FCC}}$, $\boldsymbol{c}_{\text{FCC}}$ and angles between them, α, β, γ, as a function of temperature. The I–II transition is accompanied by *ca.* 4° trigonal distortion, with virtually no change in the cell lengths. The corresponding transformation can be viewed as an elongation of a body diagonal of the cubic rotator phase I, which then becomes the trigonal phase II three-fold axis. We will refer to this particular [111] vector as $\boldsymbol{c}_{\text{trg}}$. At the II–III transition, the trigonal symmetry is lost, but two of the angles and two of the lengths remain equal (see figure 3). The transformation compatible with this geometry is a shearing of the phase II rhombohedral MD box along a [110] vector c_{bct} intersecting with c_{trg}. If α_{bct} is the [001] vector coplanar with the pair ($\boldsymbol{c}_{\text{trg}}$, $\boldsymbol{c}_{\text{bct}}$) and $\boldsymbol{b}_{\text{bct}}$, the $[1\bar{1}0]$ vector perpendicular to this plane, then a $\boldsymbol{c}_{\text{bct}}$ shear preserves the rectangular shape of the ($\boldsymbol{a}_{\text{bct}}$, $\boldsymbol{b}_{\text{bct}}$) plane. Thus, the predicted molecular centre-of-mass symmetry of phase III is monoclinic. Configurations taken from the three phases obtained in the MD calculations are shown in figure 4.

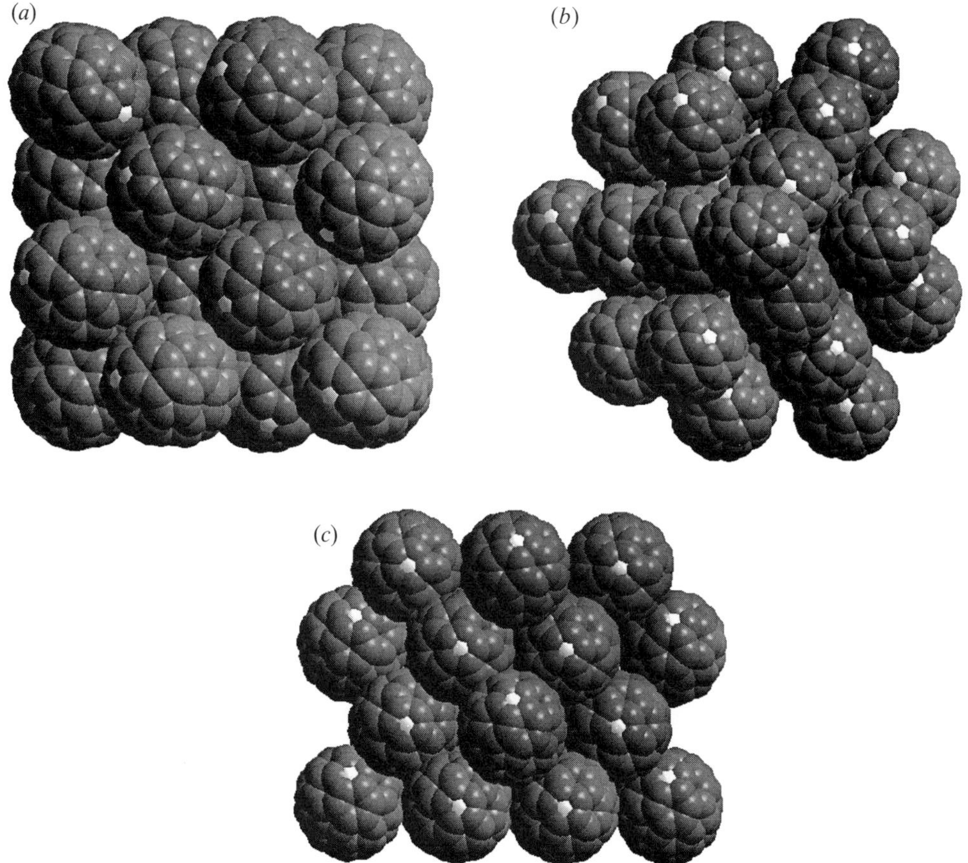

Figure 4. Instantaneous configurations taken from molecular dynamics simulations of the fullerene C_{70}. (*a*) The high-temperature ($T = 400$ K) rotator phase viewed from the $\langle 100 \rangle$ direction. (*b*) The intermediate phase viewed from the $\langle 111 \rangle$ direction. (*c*) The predicted low-temperature ($T = 10$ K) ordered phase viewed from the $\langle 110 \rangle$ direction.

With the relevant lattice directions identified, the nature of the orientational ordering can be analysed. To do so, Euler angle distributions $P(\theta), P(\phi)$ and $P(\psi)$ in the various phases have been calculated. Here, θ specifies the tilt orientation of the five-fold axis, with respect to $\boldsymbol{c}_{\text{trg}}$. The azimuthal precession angle ϕ is measured with respect to the projection of $\boldsymbol{b}_{\text{bct}}$ on the plane perpendicular to $\boldsymbol{c}_{\text{trg}}$. The body rotation (spin) angle ψ is defined with respect to a two-fold axis of the C_{70} molecule. $P(\theta)$ is not completely isotropic in phase I, but is peaked in the six equivalent $\langle 110 \rangle$ directions. In the reference frame used here, the [110] vectors are divided into two groups, three are in the $\boldsymbol{c}_{\text{trg}}$ [111] plane ($\theta = 90°$) and the other three have an inclination of $\theta = 35.3°$, with respect to $\boldsymbol{c}_{\text{trg}}$. There is a somewhat surprising tendency for five-fold axis to point in $\langle 110 \rangle$ directions rather than $\langle 111 \rangle$, as is found with an atom–atom potential model (Cheng & Klein 1992*c*). However, despite the tilt, the long axes are still aligned, on average, along the trigonal axis $\boldsymbol{c}_{\text{trg}}$. In phase II, $P(\phi)$ exhibits the same 120° modulation as in phase I, confirming that tilting towards $\langle 110 \rangle$ is preferred.

In phase III, the precessional disorder in $P(\phi)$ is removed and the competition between $\langle 111 \rangle$ and $\langle 110 \rangle$ alignment is resolved by half of the molecules orienting along the [111] vector $\boldsymbol{c}_{\text{bct}}$ ($\theta \approx 33°$, subsystem B). Simultaneously, the body axis rotation stops, at least on the MD timescale. $P(\psi)$, which was completely random in phases I and II, now splits into four peaks, each consisting of an A and B component. The AB fine structure does not show up as clear doublets because of orientational defects frozen in during the cooling process. Such an orientational glassy ground state may actually arise in the real crystal, even though it is non-cubic (Lewis & Klein 1987). However, the low-temperature structure we obtained in our simulation was sufficiently annealed to enable us to deduce an idealized phase III structure, which could then be further optimized by a steepest descent procedure.

The resulting ground state unit cell can be viewed as a monoclinic distortion of the phase I *bct* cell repeated in two of three directions, giving a unit cell containing 8 molecules and dimensions $a = 29.46$ Å, $b = 10.04$ Å, $c = 22.02$ Å, with angles $\angle\, ab = 90°$, $\angle\, bc = 90°$ and $\angle\, ac = 86.5°$. The orientations are arranged in layers AA′BB′ stacked along $\boldsymbol{c}$. The long axes of the A and A′ molecules are oriented almost perfectly in the $\boldsymbol{a}+2\boldsymbol{c}$ direction. The B and B′ molecules point along $\boldsymbol{c}$. In a given layer, the molecules are organized in rows with identical orientations parallel to $\boldsymbol{b}$ and alternating in the $\boldsymbol{a}$ direction.

TRINITY SCIENCE LIBRARY

5. Conclusions

The constant-pressure molecular dynamics technique implemented on a cluster of IBM RS/6000 workstations, operating in parallel, has enabled us to characterize possible structures of solid C_{70} unbiased by *a priori* assumptions about the orientational order.

In accordance with experimental observation and some earlier simulation results (Cheng & Klein 1992*c*) on an atom–atom model, we find orientational ordering in solid C_{70} to occur in two stages. Cooling the high-temperature phase I, produces an intermediate phase II, in which the long (five-fold) axes are oriented with a small but distinct tilt (*ca.* 18°) with respect to a unique $\langle 111\rangle$ direction. Molecular rotations about the five-fold body axis still occur. A second type of motion, which consists of 120° precessional jumps, restores the ordering $\langle 111\rangle$ direction as a threefold symmetry axis. Then, in the fully ordered phase III, half of the molecules align almost exactly along this direction $\langle 111\rangle$ and the other half along $\langle 110\rangle$. The competition between the $\langle 111\rangle$ and $\langle 110\rangle$ orientations is a consequence of a particular feature of the potential, namely repulsive interactions between C–C bonds. When the interactions are described in the simple atom–atom approximation, this curious effect is not observed and all molecules point in a common $\langle 111\rangle$ direction in both phases II and III.

The models described herein can be extended to treat the alkali metal (M) doped fullerenes M_xC_{60} (Cheng & Klein 1991*a*,*b*) and compounds, such as $C_{60}O$ (Cheng & Klein 1992*b*), but space limitations preclude any discussion of these systems. The interested reader is referred to the original articles for details.

We are grateful to Roger Larsson for installing the PVM communications software and assisting us with the parallelization of the molecular dynamics code, and François Gygi, Paul Heiney, J. E. Fischer for useful suggestions. Both A. C. and M. L. K. thank the U.S. National Science Foundation for support.

References

Allen, M. P. & Tildesley, D. J. 1987 *Computer simulation of liquids.* Oxford: Clarendon Press.

Andreoni, W., Gygi, F. & Parrinello, M. 1992 Structural and electronic properties of C_{70}. *Chem. Phys. Lett.* **189**, 241–244.

Cheng, A. & Klein, M. L. 1991*a* Molecular dynamics simulations of solid Buckminsterfullerenes. *J. phys. Chem.* **95**, 6750–6751.

Cheng, A. & Klein, M. L. 1991*b* Molecular dynamics investigation of alkali-metal-doped fullerites. *J. phys. Chem.* **95**, 9622–9625.

Cheng, A. & Klein, M. L. 1992*a* Molecular dynamics investigation of orientational freezing in solid C_{60}. *Phys. Rev.* B**45**, 1889–1895.

Cheng, A. & Klein, M. L. 1992*b* $C_{60}O$: A molecular dynamics study of rotation in the solid phase. *J. chem. Soc. Faraday Trans.* **88**, 1949–1951.

Cheng, A. & Klein, M. L. 1992*c* Solid C_{70}: A molecular dynamics study of structure and orientational ordering. *Phys. Rev.* B. **46**, 4958–4965

Copley, J. D. R., Neumann, D. A. & Cappalletti, R. L. 1992 Structure and low energy dynamics of Solid C_{60}. *Physica* B. (In the press.)

David, W. I. F., Ibberson, R. M. & Mathewman, J. C. 1991 Crystal structure and bonding of ordered C_{60}. *Nature, Lond.* **353**, 147–149.

David, W. I. F., Ibberson, R. M., Dennis, T. J. S., Hare, J. P. & Prassides, K. 1992 Structural phase transitions in the fullerene C_{60}. *Eur. Phys. Lett.* **18**, 219–225.

Dworkin, A., Szwarc, H. & Leach, S. 1991 Mise en évidence thermodynamique d'une transition de phase dans le fulleréne C_{60} cristallin. *C.r. Acad. Sci., Paris* II **312**, 979–982.

Feuston, B. P., Andreoni, W., Parrinello, M. & Clementi, E. 1991 Electronic and vibrational properties of C_{60} at finite temperature from *ab initio* molecular dynamics. *Phys. Rev.* B **44**, 4056–4059.

Fowler, P. W., Lazzeretti, P. & Zanasi, R. 1990 Electric and magnetic properties of the aromatic sixty-carbon cage. *Chem. Phys. Lett.* **165**, 79–86.

Fowler, P. W., Lazzeretti, P., Malagoli, M. & Zanasi, R. 1991 Magnetic properties of C_{60} and C_{70}. *Chem. Phys. Lett.* **179**, 174–180.

Guo, Y., Karasawa, N. & Goddard III, W. A. 1991 Prediction of fullerene packing in C_{60} and C_{70} crystals. *Nature, Lond.* **351**, 464–467.

Heiney, P. A. *et al.* 1991 Orientational phase transition in solid C_{60}. *Phys. Rev. Lett.* **66**, 2911–1468.

Heiney, P. A. *et al.* 1992 Discontinuous volume change at the orientational-ordering transition in solid C_{60}. *Phys. Rev.* B**45**, 4544–4547.

Johnson, R. D., Meijer, G., Salem, J. R. & Bethune, D. S. 1991 Two-dimensional nuclear resonance study of the structure of the fullerene C_{70}. *J. Am. chem. Soc.* **113**, 3619–3621.

Johnson, R. D., Yannoni, C. S. & Dorn, H. C. 1992 C_{60} rotation in the solid state: Dynamics of a faceted spherical top. *Science, Wash.* **255**, 1235–1238.

Kriza, G. *et al.* 1991 Pressure dependence of the structural phase transition in C_{60}. *J. Phys. I* **1**, 1361–1364.

Lewis, L. J. & Klein, M. L. 1987 Is the ground state of (KCl)(KCN) a non-cubic orientational glass? *Phys. Rev. Lett.* **59**, 1837–1840.

Lu, J. P., Li, X.-P. & Martin, R. M. 1992 Ground state and phase transitions in solid C_{60}. *Phys. Rev. Lett.* **68**, 1551–1554.

Michel, K. H., Copley, J. D. R. & Neumann, D. A. 1992 Molecular theory of orientational disorder and the orientational phase transition in solid C_{60}. *Phys. Rev. Lett.* **68**, 2929–2932.

Neumann, D. A. *et al.* 1991 Coherent quasielastic neutron scattering study of rotational dynamics of C_{60} in the orientationally disordered phase. *Phys. Rev. Lett.* **67**, 3808–3811.

Nosé, S. & Klein, M. L. 1983 Constant pressure molecular dynamics for molecular systems. *Molec. Phys.* **50**, 1055–1076.

Pan, C., Sampson, M. P., Chai, Y., Hauge, R. H. & Margrave, J. L. 1991 Heats of sublimation from a polycrystalline mixture of C_{60} and C_{70}. *J. phys. Chem.* **95**, 2944–2946.

Parrinello, M. & Rahman, A. 1980 A new constant pressure molecular dynamics. *Phys. Rev. Lett.* **45**, 1196–1199.

Raich, J. & Mills, R. L. 1971 α–γ Transition in solid nitrogen and carbon monoxide at high pressure. *J. chem. Phys.* **55**, 1811–1187.

Sachidanandum, R. & Harris, A. B. 1991 Comment on: Orientational phase transition in solid C_{60}. *Phys. Rev. Lett.* **67**, 1467.

Shi, X. D., Kortan, A. R., Williams, J. M., Kini, A. M., Savall, B. M. & Chaiken, P. M. 1992 Sound velocity and attenuation in single-crystal C_{60}. *Phys. Rev. Lett.* **68**, 827–830.

Sprik, M., Cheng, A. & Klein, M. L. 1992 Modelling the orientational ordering transition in solid C_{60}. *J. phys. Chem.* **96**, 2027–2029.

Taylor, R., Hare, J. P., Abdul-Sada, A. K. & Kroto, H. W. 1990 Isolation, separation and characterization of the fullerenes C_{60} and C_{70}: The third form of carbon. *J. chem. Soc. chem. Commun.* 1423–1425.

Tycko, R., Dabbagh, R. M., Flemming, R. M., Haddon, R. C., Makhija, A. V. & Zahurak, S. M. 1991*a* Molecular dynamics and phase transition in solid C_{60}. *Phys. Rev. Lett.* **67**, 1886–1889.

Tycko, R. *et al.* 1991*b* Solid-state magnetic resonance spectroscopy of fullerenes. *J. phys. Chem.* **95**, 518–520.

van Loosdrecht, P. H. M., van Bentum, P. J. M. & Meijer, G. 1992 Rotational ordering transition in single-crystal C_{60} studied by Raman spectroscopy. *Phys. Rev. Lett.* **68**, 1176–1179.

Vaughan, G. B. M. *et al.* 1992 Orientational disorder in solvent-free solid C_{70}. *Science, Wash.* **254**, 1350–1353.

Yannoni, C. S., Johnson, R. D., Meijer, G., Bethune, D. S. & Salem, J. R. 1991 *J. phys. Chem.* **95**, 9–10.

Discussion

M. Lal (*Unilever Research, Port Sunlight Laboratory, U.K.*): The maximum time covered in your MD simulations is 200 ps. How did you establish that the water molecules at the monolayer of the hydrocarbon chain molecules have attained the equilibrium state in the droplet form in coexistence with vapour? If the initial water layer was several molecules thick, would you then expect that the equilibrium state would be the one in which a residual monolayer of water molecules exists in equilibrium with the droplet?

M. L. Klein: The equilibrium state was checked in the usual fashion. For example, we monitor the height of the droplet's centre of mass as a function of time, the radial density profile, energy, etc. We are confident that our results (J. Hautman & M. L. Klein, *Phys. Rev. Lett.* **67**, 1763–1766 (1991)) are indeed correct.

Simulations of materials: from electrons to friction

By Uzi Landman, R. N. Barnett and W. D. Luedtke
School of Physics, Georgia Institute of Technology, Atlanta, Georgia 30332, U.S.A.

Quantum and classical molecular dynamics simulations are discussed, illustrating the applicability of computer-based modelling to a broad range of materials systems and phenomena. Case studies discussed include: quantum simulations of fission dynamics of charged atomic clusters and metallization of finite, small, alkali-halide crystals, classical molecular dynamics investigations of the consequences of interfacial adhesive interactions leading to the formation of intermetallic junctions, and the molecular mechanisms of capillary processes.

1. Introduction

Traditionally, modes of scientific investigations are classified, based on thematic, methodological or historical arguments, as theoretical or experimental in nature. The emergence of computers as high-powered research tools, and the development and proliferation of computer-based modelling and simulations, add a new dimension to our capabilities to explore natural phenomena and broaden our perspectives. Furthermore, by their nature computer-based modelling and simulations of materials are akin to theoretical experiments, where the physical system is represented by a theoretical model and a numerical algorithm, computational code, a computer, and various peripheral instruments (such as visualization devices) serve as the laboratory.

Computer simulations, where the evolution of a physical system is simulated, with refined temporal and spatial resolution, via a direct numerical solution of the equations of motion (quantum or classical) open new avenues in investigations of the microscopic origins of material phenomena (see articles in Catlow & Mackrodt 1982; Landau *et al.* 1988; Vitek & Srolovitz 1988; Nieminen *et al.* 1990). These methods alleviate certain of the major difficulties that hamper other theoretical approaches, particularly for complex systems such as those characterized by a large number of degrees of freedom, lack of symmetry, nonlinearities and complicated interactions. In addition to comparisons with experimental data, computer simulations can be used as a source of physical information, which is not accessible to laboratory experiments, and in some instances the computer experiment itself serves as a testing ground for theoretical concepts. Consequently, computer-based theoretical investigations of materials systems allow theorists to make contributions in explaining and elucidating the results of specific experiments, guide the developments of concepts, principles, and theories unifying a range of observations, and predict new behaviour.

Along with the opportunities and advantages presented by computer simulations, such studies do have certain limitations (mostly of a technical nature) and pitfalls associated with them. The main issues involved in critical assessment of simulation studies are: (i) the faithfulness of the simulation model, focusing mainly on our knowledge of the interaction potentials; (ii) the spatial dimension, i.e. finite size of the computational cell and imposed periodic boundary conditions in the case of

Phil. Trans. R. Soc. Lond. A (1992) **341**, 337–350
Printed in Great Britain
© 1992 The Royal Society

simulations of extended systems; and (iii) the finite time span of the simulation. The physical size and time extent used in a simulation are constrained by the available computational resources, and their adequacy depends upon characteristics of the specific system and phenomenon being studied, such as the state of aggregation, range of interactions, characteristic relaxation times, the scale of intrinsic spatial and temporal fluctuations and ambient conditions. The symbiotic relationship between large-scale computer-based modelling and computer technology benefits both areas. Indeed, in recent years the quest to simulate larger and more complex systems influenced the development of new computational strategies and computer architectures, and vice versa.

Basic understanding of the structure and dynamics of materials and their properties often requires knowledge on a microscopic level of the underlying energetics and interaction mechanisms, whose consequences we observe and measure. In this paper we discuss and demonstrate, using several case studies and various simulation techniques ranging from quantum to classical molecular dynamics applied to finite aggregates and extended systems, the wealth of information that can be obtained via quantum and classical simulations of materials, and the insights that such studies provide pertaining to fundamental as well as technology-related scientific issues.

2. Case studies

(a) *Quantum molecular dynamics: clusters*

Among the principal motivations for studies of clusters are a large number of observations that finite materials aggregates exhibit size-dependent physical and chemical properties, and the expectation that exploration of the systematics of such size effects and of their physical origins (both in the classical and quantum mechanical domains), would open avenues for understanding and elucidation of the size evolutionary patterns of materials properties, from the atomic and molecular scale to the condensed phase régime (see articles in Jena *et al.* 1987; Benedek *et al.* 1988).

(i) *Patterns and barriers for fission of charged small metal clusters*

Recent studies of metallic clusters (particularly of simple metals) unveiled systematic energetic, stability, spectral and fragmentation trends (see Barnett *et al.* 1991, and references therein), which bear close analogy to corresponding phenomena exhibited by atomic nuclei, suggesting an intriguing universality of the physical behaviour of finite size aggregates, though governed by interactions of differing spatial and energy scales.

To investigate energetic patterns and dynamics of fission of small Na_n^{+2} ($n \leqslant 12$) clusters, we have used our newly developed simulation method that combines classical molecular dynamics, or energy minimization, on the Born–Oppenheimer (BO) ground-state potential surface, with electronic structure calculations via the Kohn–Sham (KS) formulation of the local spin-density (LSD) functional method (Barnett *et al.* 1991; Rajagopal *et al* 1991). In dynamical simulations the ionic (classical) degrees of freedom evolve on the electronic BO surface, which is calculated after each classical step. Minimum energy structures were obtained by a steepest-descent method, starting from configurations selected from finite temperature simulations. Barrier heights and shapes were obtained by constrained energy minimization, with the centre-of-mass distance between the fragments specified.

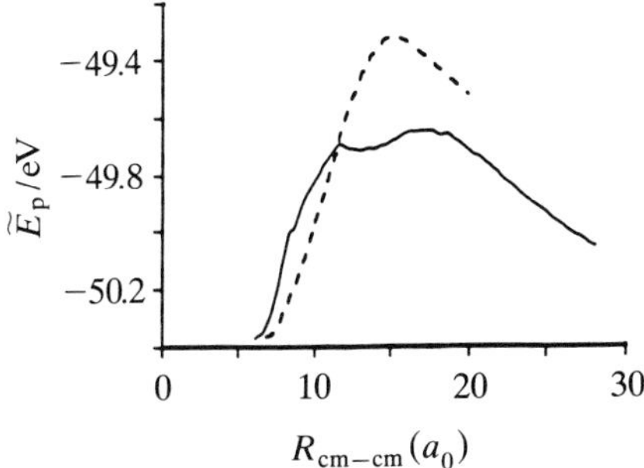

Figure 1. Potential energy against distance between the centre of mass for the fission $Na_{10}^{+2} \rightarrow Na_7^+ + Na_3^+$ (solid) and $Na_{10}^{+2} \rightarrow Na_9^+ + Na^+$ (dashed), obtained via constrained minimization of the LSD ground-state energy of the system.

Examination of the energetics of the clusters and of the various fragmentation channels reveals that in all cases the energetically favoured channel is $Na_n^{+2} \rightarrow Na_{n-3}^+ + Na_3^+$ ($n \leqslant 12$), i.e. asymmetric fission (except for $n = 6$), associated with shell closing in the product Na_3^+ cluster. Furthermore, our calculations show that for $n > 6$ fission involves energy barriers. The barriers for $n = 8$, 10 and 12 have been determined via constrained minimization to be: 0.16 eV, 0.71 eV and 0.29 eV for the energetically favoured channel, and larger barriers were found for the ejection of Na^+ from these clusters (0.43 eV and 1.03 eV for $n = 8$ and 10 respectively). The fission barriers for Na_{10}^{+2} are higher because of the closed-shell structure of this parent cluster.

The potential energies along the reaction coordinates for the energetically favoured channel and for Na^+ ejection, in the case of Na_{10}^{+2} are shown in figure 1. The most interesting feature seen from the figure is the rather unusual shape of the barrier for the favoured fission channel (also found for the asymmetric fragmentation of Na_{12}^{+2}). Although double-hump barriers have been long discussed in the theory of nuclear fission (Preston & Bhaduri 1975), to our knowledge this is the first time that they have been calculated in the context of asymmetric fission of charged atomic clusters (Barnett *et al.* 1991).

The existence of the double-humped barrier is reflected in the dynamics of the fission process of Na_{10}^{+2} displayed in figure 2, illustrating also the new capabilities afforded via quantum simulations for explorations of the dynamics of physical and chemical phenomena. This simulation started from a 600 K Na_{10} cluster from which two electrons were removed (requiring 11.23 eV) and 0.77 eV was added to the classical ionic kinetic energy. The variation of the centre-of-mass distance with time (figure 2*a*) exhibits a plateau for 750 fs $\leqslant t \leqslant$ 2000 fs (see also the behaviour of the electronic contribution to the potential energy of the system against time in figure 2*c*). The contours of the electronic charge density of the system (figure 2*d–f*), at selected times, and the corresponding cluster configurations (figure 2*g–i*), reveal that the fission process involves a precursor state that undergoes a structural isomerization before the eventual separation of the Na_7^+ and Na_3^+ fission products. In this context we remark that examination of the contributions of individual Kohn–Sham orbitals to the total density for the intermediate stage (figure 2*e*) reveals that the lowest-energy orbital (s-like) is localized on the Na_7^+ fragment, the second orbital is s-like localized on Na_3^+, the third is a p-like bonding orbital distributed over the two fragments, and the highest orbital is localized on the larger fragment.

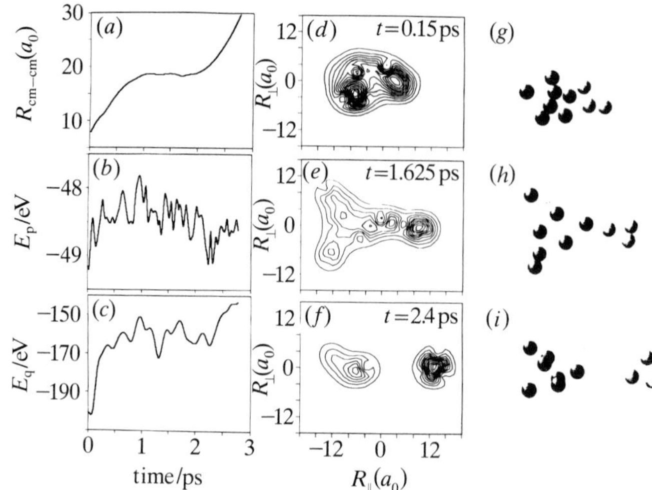

Figure 2. Fragmentation dynamics of Na_{10}^{+2}. (*a*)–(*c*) R_{cm-cm} between the fission products, total potential energy (E_p), and the electronic contribution (E_q) against time. (*d*)–(*f*) Contours of the total electronic charge distribution, calculated in the plane containing the two centres of masses. The R axis is parallel to R_{cm-cm}. (*g*)–(*i*) Ionic configurations for the times given in (*d*)–(*f*). Dark and light spheres represent ions in the large and small fragments respectively. Energy, distance, and time in units of eV, Bohr (a_o) and ps respectively.

(ii) *Metallization patterns of ionic clusters*

Investigations of localization modes, structure, dynamics and spectra of alkali-halide clusters containing multiple excess electrons (i.e. those electrons which substitute F^- anions in Na_nF_m, $m < n$) open new dimensions for studies of excess-electron localization and bonding in ionic clusters (Landman *et al.* 1985; Scharf *et al.* 1987; Rajagopal *et al.* 1990, 1991; for studies of modes and dynamics of electron and dielectron localization in water clusters see Barnett *et al.* 1990; Kaukonen *et al.* 1992), since the process of metallization of an initially stoichiometric ionic cluster (i.e. starting from an ionic cluster with $n = m$, and successively substituting anions by electrons, ultimately resulting in a neutral metallic cluster Na_n) may portray a transition from an insulating to a metallic state in a finite system. In this context we note that while transitions from F-centre to metallic behaviour have been observed in bulk molten alkali halides at high excess-metal concentrations, experimental data on metal-rich alkali-halide clusters are preliminary in nature (see references in Rajagopal *et al.* 1991).

In our BO–LSD simulations the inter-ionic interactions are described by the Born and Huang parametrized potentials which we have tested in previous studies, the interaction between the electrons and Na^+ is given by a norm-conserving pseudopotential, and a Coulomb repulsive potential describes the interaction between the electrons and F^- (Rajagopal *et al.* 1991).

We begin with a metallization sequence (MS) of a small cluster, Na_4F_m ($0 \leqslant m \leqslant 4$), which for $m = 4$ is a stable ionic cuboid and at the other extreme ($m = 0$) is a stable planar rhombohedral Na_4 cluster (see figure 3*a*–*e*). The structures exhibited by this MS demonstrate a systematic trend: structures of neutral small clusters belonging to a MS are of the same dimensionality and symmetry of the corresponding parent cluster, converting to the structure of the corresponding metal cluster upon complete metallization ($m = 0$). Furthermore, the excess electrons in these clusters,

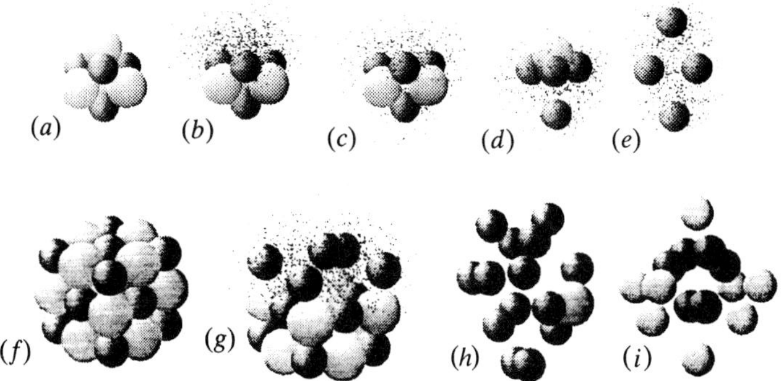

Figure 3. Optimal structures of Na_nF_m clusters. (*a*)–(*e*) MS for Na_4F_m ($0 \leqslant m \leqslant 4$). (*f*)–(*i*) Structures of $Na_{14}F_m$, for $m = 13$, 9, 1 and 0 respectively. Large and small spheres denote F^- and Na^+. Small dots represent the total excess electronic distribution.

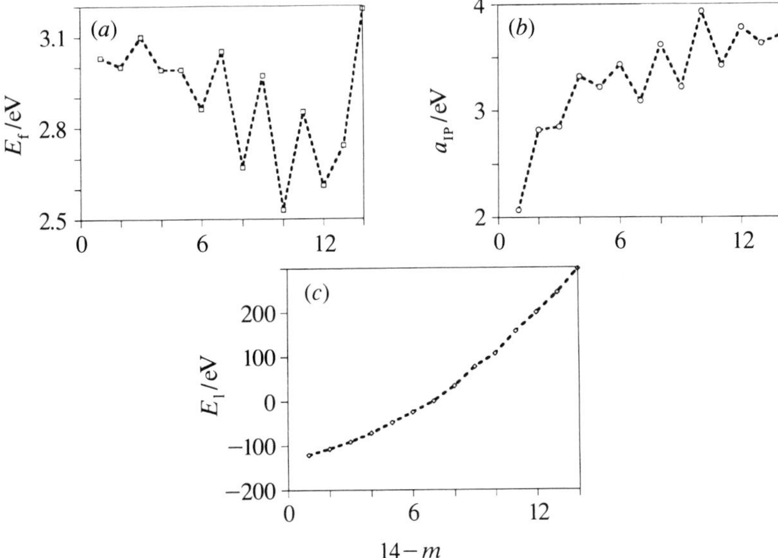

Figure 4. Energetics in the MS for $Na_{14}F_m$ against $14-m$, ($13 \geqslant m \geqslant 0$). Formation, adiabatic ionization, and ion-ion interaction energies in (*a*)–(*c*) respectively.

substituting for halide anions, maintain the cohesion and structural integrity of the clusters and their distributions exhibit a delocalized, metallic character (with the regions occupied by the remaining anions, i.e. for $0 < m < n$, excluded).

The MS for a larger cluster $Na_{14}F_m$ ($0 \leqslant m \leqslant 14$), whose $Na_{14}F_{13}$ member is a particularly stable cluster ($Na_{14}F_{13}^+$ is a 'magic number' ionic cluster, i.e. a $3 \times 3 \times 3$ filled cuboid structure, with three ions of alternating charges on an edge), exhibits another novel result: face (or atomic layer) metallization (segregation), as seen in figure 3*g*. The energetically optimal metallization sequence of this cluster proceeds via successive removal of neighbouring halogens from one face of the cluster, resulting (for $Na_{14}F_9$) in a segregated metal layer.

Examination of the pattern of metallization, and in particular the MS for $Na_{14}F_m$ (partly displayed in figure 3), reveals several trends. (i) The F-centre formation energies (E_f) (i.e. the difference between the total energies of Na_nF_{m-1} and Na_nF_m)

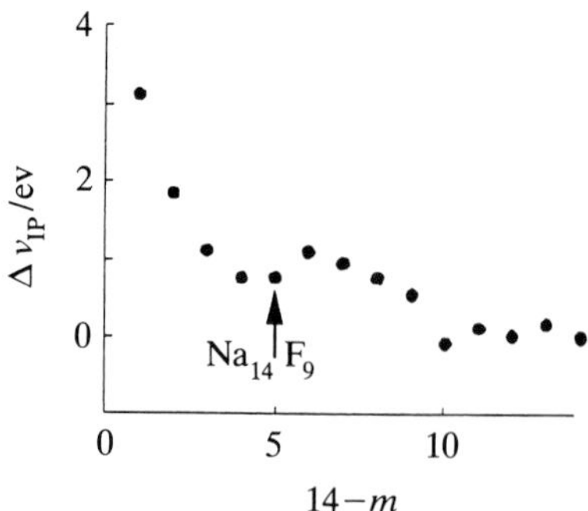

Figure 5. Δv_{IP} against $14-m$, ($13 \geqslant m \geqslant 0$), for the MS of $Na_{14}F_m$. Note local minimum corresponding to $Na_{14}F_9$ and small values for $m \leqslant 4$, indicating enhanced metallic character.

are between 2.5 to 3.1 eV (figure 4). It is of interest that within a MS the formation energies, as well as the adiabatic (and vertical) ionization potentials, exhibit odd-even oscillations in $n-m$ indicating that perhaps these excess-metal systems may be regarded as composed of a 'metallic' component and a molecular-ionic one, symbolically represented as $Na_nF_m = Na_{n-m}(NaF)_m$. Our results (structural and energetic) for the bare metal and mixed clusters are in good agreement with experimental data and previous calculations, when available (Bonacic-Koutecky *et al.* 1991). (ii) A particularly interesting trend is exhibited by $\Delta v_{IP} = v_{IP}[Na_{n-m}] - v_{IP}[Na_{n-m}(NaF)_m]$, plotted in figure 5 against $n-m$ ($n = 14$ and $m = 1, 2, 3, \ldots$), which expresses the difference between the vertical ionization potential of a bare sodium cluster containing $n-m$ atoms and that of the mixed $Na_nF_m \equiv Na_{n-m}(NaF)_m$ cluster. Δv_{IP} may be regarded as a measure of metallic behaviour in the mixed cluster. As seen, for $n-m = 5$ (i.e. corresponding to $Na_{14}F_9$) Δv_{IP} attains a local minimum, suggesting enhanced metallic behaviour for the face-segregated cluster. Moreover, for $n-m \geqslant 10$ (i.e. $m \leqslant 4$), the perturbing effect of the ionic component of the cluster is small, and the electronic properties of the clusters approach those of the corresponding bare metal ones.

(b) *Adhesive junctions and capillary columns*

Understanding the atomistic mechanisms, energetics, structure and dynamics underlying the interactions and physical processes that occur when two materials are brought together (or separated) is fundamentally important to basic and applied problems such as adhesion, capillarity, contact formation, surface deformations, materials elastic and plastic response characteristics, materials hardness, micro- and nano-indentation, friction, lubrication, and wear, fracture and atomic-scale probing, modifications and manipulations of materials surfaces (see references in Landman *et al.* 1990, 1992; Landman & Luedtke 1991, 1992).

(i) *Adhesive intermetallic interfaces*

The first to introduce the notion of adhesive forces between material bodies in contact, and their contribution to the overall frictional resistance experienced by sliding bodies, was Desaguliers (1734, see also Dowson 1979), whose ideas on friction were conceived in the context of the role of surface finish, where he writes: '... the flat surfaces of metals or other Bodies may be so far polish'd as to increase Friction and this is a mechanical Paradox: but the reason will appear when we consider that the

Attraction of Cohesion becomes sensible as we bring the Surfaces of Bodies nearer and nearer to Contact'. These observations are incorporated in the current view that friction is the force required to shear intermetallic junctions plus the force required to plough the surface of the softer metal by the asperities on the harder surface (Bowden & Tabor 1973).

Application of newly developed theoretical and experimental (tip-based and surface-force apparatus) techniques (Israelachvili 1992; Sarid 1992; Murday & Colton 1990; Behm *et al.* 1989) promises to provide significant insights concerning the microscopic mechanisms and the role of surface forces in the formation of microcontacts and to enhance our understanding of fundamental issues pertaining to interfacial adherence, microindentation, surface deformations and the transition from elastic to elastoplastic or fully developed plastic response of materials, thin films, coatings, wetting and lubrication. Additionally, such studies in conjunction with atomistic investigations allow critical assessment of the range of validity of continuum-based theories of these phenomena and can inspire improved analytical formulations. Finally, knowledge of the interactions and atomic-scale processes occurring between small tips and materials surfaces, and their consequences, is of crucial importance to optimize, control, interpret and design experiments using novel tip-based microscopies.

As an illustration of recent progress in this area we show in figure 6*a* (solid line) the force between a nickel tip and Au(001) surface as they approach each other and are subsequently separated (Landman *et al.* 1990). In these large-scale molecular dynamical (MD) simulations the EAM potentials (Adams *et al.* 1989) are used for a system consisting of eight dynamic and three static Au(001) layers, with 450 atoms per layer, and the Ni tip consists of a bottom layer of 72 atoms exposing a (001) facet, the next layer consists of 128 atoms and the remaining six layers contain 200 dynamic atoms each. This gives the tip an effective radius of curvature of about 30 Å†. The static holder of the tip (rigid cantilever) consists of 1176 atoms arranged in three (001) layers. All simulations were performed at 300 K, and following equilibration motion of the tip in the normal direction relative to the substrate occurs via changing the position of the tip-holder assembly in increments of 0.25 Å, and after each increment fully relaxing the system, i.e. dynamically evolving until no discernable variations in system properties are observed beyond natural fluctuations.

The main features observed from figure 6*a* (and in accompanying atomic-force microscopy experiments (Landman *et al.* 1990)) are the following.

1. The onset of an instability, signified by a sharp increase in the attraction between the tip and the substrate, occurring at a distance $d_{ts} \approx 4.20$ Å, between the bottom-most atomic layer of the tip and the top-most atomic layer of the gold surface. This instability, which results in a jump-to-contact (JC) phenomenon, occurs via a plastic deformation of the Au(001) surface in the region under the tip, involving displacement of atoms in that region by about 2 Å in 1 ps (after the JC occurs the spacing between the bottom layer of the Ni tip and the adherent (wetting) layer of Au atoms decreases to 2.1 Å from a value of 4.2 Å before JC).

The jump-to-contact phenomenon in metallic systems is driven by the marked tendency of the atoms at the interfacial regions of the tip and substrate materials to optimize their embedding energies (which are density dependent, deriving from the tails of the atomic electronic charge densities) while maintaining their individual

† 1 Å = 10^{-10} m = 10^{-1} nm.

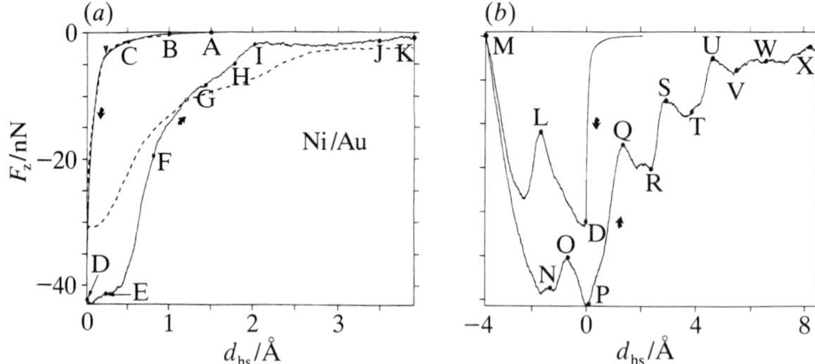

Figure 6. Calculated force on the tip atoms, F_z, against tip-to-sample distance d_{hs}, between a Ni tip and an Au sample for: (*a*) approach and jump-to-contact followed by separation; (*b*) approach, jump-to-contact, indentation, and subsequent separation; d_{hs} denotes the distance between the rigid tip-holder assembly and the static substrate of the Au surface ($d_{hs} = 0$ at the jump-to-contact point marked D). The capital letters on the curves denote the actual distances, d_{ts}, between the bottom part of the Ni tip and the Au surface; in (*a*) A = 5.7 Å, B = 5.2 Å, C = 4.7 Å, D = 3.8 Å, E = 4.4 Å, F = 4.85 Å, G = 5.5 Å, H = 5.9 Å, I = 6.2 Å, J = 7.5 Å and K = 8.0 Å; in (*b*) D = 3.8 Å, L = 2.4, M = 0.8 Å, N = 2.6 Å, O = 3.0 Å, P = 3.8 Å, Q = 5.4 Å, R = 6.4 Å, S = 7.0 Å, T = 7.7 Å, U = 9.1 Å, V = 9.6 Å, W = 10.5 Å, and X = 12.8 Å. Forces in units of nanonewtons, and distance in ångströms.

material cohesive binding (in the Ni and Au) albeit strained due to the deformation caused by the atomic displacements during the JC process. In this context we note the difference between the surface energies of the two metals, with the one for Ni markedly larger than that of Au, and the differences in their mechanical properties (for example, the elastic moduli are 21×10^{10} and 8.2×10^{10} N m^{-2} for Ni and Au respectively).

Additional insight into the JC process is provided by the local hydrostatic pressure and stress distribution in the materials (evaluated from the atomic stress tensors) after contact formation. Both the structural deformation profile of the system and the stress distribution which we find in our atomistic MD simulations are similar, in general terms, to those described by certain modern contact mechanics theories (see review in Israelachvili 1992; Burnham *et al.* 1991; articles in Pollock & Singer 1992; see also references in Landman *et al.* 1990), where the influence of adhesive interactions is included, indicating the applicability of such continuum descriptions even to systems characterized by rather small spatial dimensions.

2. Reversal of the direction of the tip motion relative to the substrate starting from the point of adhesive contact (point D in figure 6*a*) results in a marked hysteresis (we note that separating the two before contact results in no hysteresis). The hysteresis is a consequence of the adhesive bonding between the two materials, which upon retraction of the tip results in the formation of a connective neck, with a monolayer of gold atoms coating (wetting) the retracted Ni tip.

It is of interest to note that repeating the process but using the gold coated nickel tip, results again in a JC instability (though of reduced magnitude) and hysteresis upon retraction (see figure 6*a*, dashed line). However, while plastic deformation of the surface and neck formation were observed during the lift-off process, no gold atom transfer to the coated tip was observed (Landman *et al.* 1992).

Allowing the tip to advance past the jump-to-contact point, that is indenting the surface, results in the force against distance curve recorded in figure 6*b*. Of particular

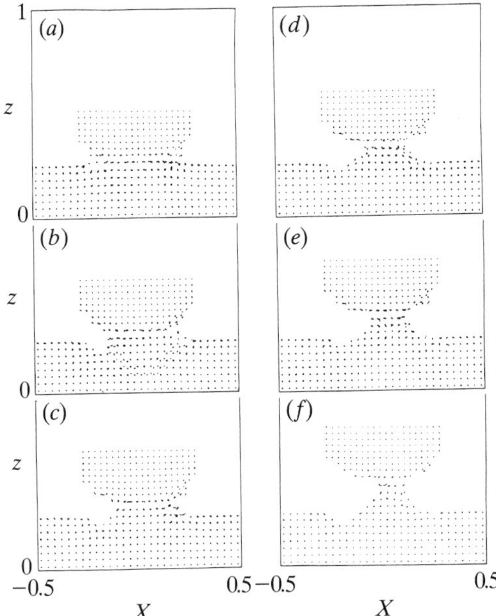

Figure 7. Atomic configurations in slices through the system illustrating the formation of a connective neck between the Ni tip and the Au substrate during separation following indentation. The Ni tip occupies the topmost eight layers. The configurations (*a*)–(*f*) correspond to the stages marked O, Q, S, U, W, and X in figure 6*b*. Note the crystalline structure of the neck. Successive elongations of the neck, upon increased separation between the tip-holder assembly and the substrate, occur via structural transformation resulting in successive addition of layers in the neck accompanied by narrowing (i.e. reduction in cross-sectional area of the neck). Distance in units of X and Z, with $X = 1$ and $Z = 1$ corresponding to 61.2 Å.

interest are the oscillations in the force upon retraction from the point of zero force (point M in figure 6*b*), associated with the formation and elongation of a connective junction consisting mainly of gold atoms (see figure 7). Further insight into the microscopic mechanism of elongation of the connective neck can be gained via consideration of the variation of the second invariant of the stress deviator, J_2, which is proportional to the stored strain energy and is related to the von Mises shear strain-energy criterion for the onset of plastic yielding (Dieter 1967). Such analysis reveals that between each of the elongation events (i.e. layer additions, points marked Q, S, U, W and X) the initial response of the system to the strain induced by the increased separation is mainly elastic (segments OP, QR, ST, UV in figure 6*b*) accompanied by a gradual increase of the stored strain energy. The onsets of the stages of structural rearrangements are found to be correlated with a critical maximum value of $\sqrt{J_2}$ of about 3 GPa (occurring for states at the end of the intervals marked OP, QR, ST and UV in figure 6*b*), localized in the neck in the region of the ensuing structural transformation. After each of the elongation events the maximum value of $\sqrt{J_2}$ (for the states marked Q, S, U, W and X in figure 6*b*) drops to approximately 2 GPa.

The theoretically predicted increased hysteresis upon tip–substrate separation following indentation, relative to that found after contact (compare figure 6*a*, *b*), is also observed experimentally (Landman *et al.* 1990). While in the original experiments the non-monotonic features found in the simulations (figure 6*b*) were not discernible in the force against distance data, they have been observed in recent experiments performed under ultra-high vacuum conditions (C. M. Mate, personal communication).

(ii) *Capillary junctions*

Our discussion up to this point was confined to the interaction between material tips and bare crystalline substrates. Motivated by the fundamental and practical importance of understanding the properties of adsorbed molecularly thin films and phenomena occurring when films are confined between two solid surfaces, pertaining to diverse fields such as fluid-film lubrication, prevention of degradation and wear, wetting, spreading and drainage, the mechanical response and relaxation of adsorbed organic films, and AFM measurements, we have initiated most recently investigations of such systems (Landman *et al.* 1992; Landman & Luedtke 1992). Among the issues that we attempt to address are the structure, dynamics, and response of confined complex films, their rheological properties, and modification which they may cause to adhesive and tribological phenomena, such as inhibition of jump-to-contact instabilities and prevention of contact junction formation.

The molecular films that we studied, *n*-hexadecane ($C_{16}H_{34}$), were modelled by interaction potentials developed by Ryckaert & Bellemans (1978), which have been used before in investigations of the thermodynamic, structural and rheological properties of bulk liquid *n*-alkanes, and adsorbed hexadecane films of variable thickness (butane and decane (Leggeter & Tildesley 1989; Xia *et al.* 1992; Ribarsky & Landman 1992)). The metallic substrate (Au) and tip (Ni) were modeled using the EAM potentials and the interactions between the *n*-hexadecane molecules and the metallic tip and substrate were described using LJ potentials, determined by fitting to experimentally estimated adsorption energies (Xia *et al.* 1992; Landman & Luedtke 1992).

Studies of the response of the system revealed (Landman *et al.* 1992) that lowering of a faceted nickel tip towards a gold surface covered by a thin adsorbed *n*-hexadecane film results first in small attraction between the film and the tip followed, upon further lowering of the tip, by ordering (layering) of the molecular film (see below). During continued approach of the tip toward the surface the total interaction between the tip and the substrate (metal plus film) is repulsive, and the process is accompanied by molecular drainage from the region directly under the tip, wetting of the sides of the tip, and ordering of the adsorbed molecular monolayer under the tip. Further lowering of the tip is accompanied by inward deformation of the substrate and eventual formation of intermetallic contact (occurring via displacement of surface gold atoms towards the tip) which is accompanied by partial molecular drainage and results in a net attractive force on the tip. We note that, unlike the case of the bare metal interface discussed before, formation of contact in the presence of an adsorbed film required the application of a relatively high load.

The last case study we discuss illustrates the consequences of tip–sample interactions in investigations of capillary phenomena occurring upon approach and subsequent retraction of a blunted tip, to and from a liquid alkane film (Landman & Luedtke 1992). In this study the simulated system is composed of a large gold static substrate, a static nickel tip, and a film of alkane molecules whose number is controlled such as to approximately conserve the chemical potential of the system (i.e. a pseudo grand-canonical molecular dynamics simulation where the thickness of the adsorbed molecular liquid film in regions away from the tip is monitored

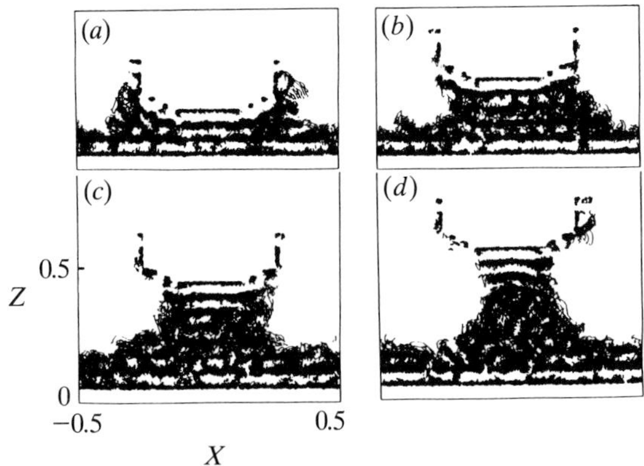

Figure 8. Short-time trajectories, obtained via MD simulations at 350 K, of hexadecane molecules forming a capillary column between a gold substrate and a nickel tip at four stages of the tip-lifting process (trajectories were plotted in a 23 Å wide slice through the middle of the system). The distances between the tip and the substrate for the configurations shown in (*a*)–(*d*) are $d_{ts} = 19.3$ Å, 28.1 Å, 36.9 Å and 45.7 Å, correspondingly. The length scale of the calculational cell, which was periodically repeated along the x and y directions is 77.5 Å.

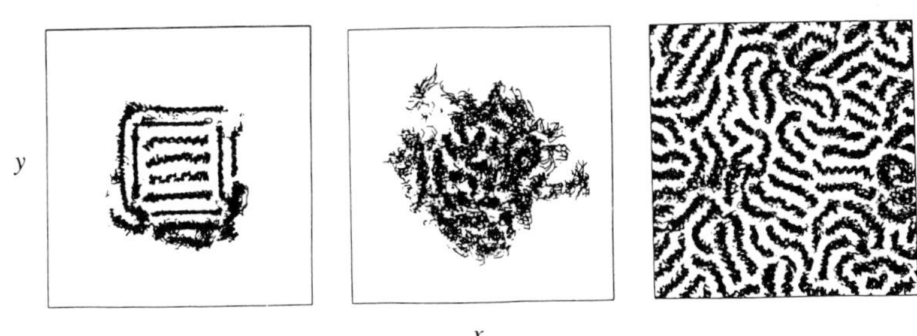

Figure 9. Short-time trajectories of hexadecane molecules in three regions of the capillary column formed between a gold surface and a nickel tip separated by 36.9 Å (corresponding to the configuration shown in figure 8*c*). The bottom configuration corresponding to the region of the film closest to the gold surface illustrates preferential orientation of the molecules parallel to the surface and islands of intermolecular ordering. The middle panel illustrates lack of order in the middle of the liquid junction, while the molecular configuration shown at top, corresponding to a region closest to the bottom of the nickel tip, illustrates a high degree of preferential parallel molecular orientation and intermolecular order. Length scale as in figure 8.

throughout the wetting and growth of the capillary liquid column, and molecules are added at the edges of the edges of the computational cell to compensate for those that were pulled up toward the tip by the adhesive and capillary forces).

Four configurations of the alkane film are shown in figure 8, each corresponding to an equilibrium state of the system at a given separation between the tip and the surface. The configurations shown in figure 8*a*, *b* correspond to distances $d_{ts} = 19.3$ Å and $d_{ts} = 28.1$ Å between the bottom layer of the Ni tip and the topmost layer of the gold surface. As seen in these stages the liquid film in the region under the tip consists of four and six distinct layers, respectively, while the arrangement of molecules on the sides of the liquid-junction is less ordered, receding to a two-layer adsorbed film further away from the tip. At this point it is of interest to note that the degree of layering is found to exhibit a marked sensitivity to the distance d_{ts} between the tip and the substrate, maximizing when d_{ts} is commensurate with an integral number of

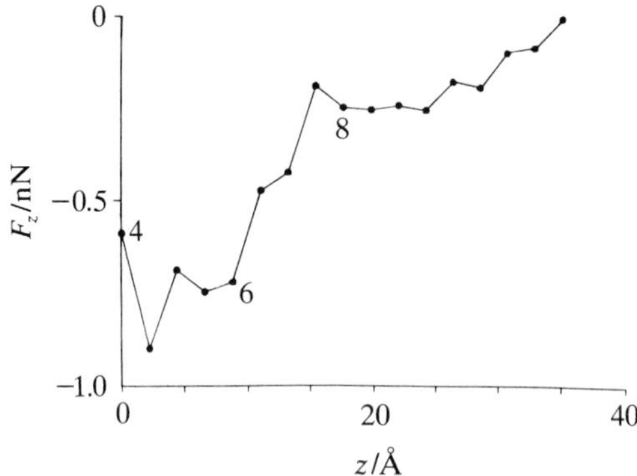

Figure 10. Normal force on the nickel tip (in nN) against the distance between the tip and the gold surface, Z (in Å) with the origin taken as that corresponding to a four-layer capillary column (see figure 8*a*). Dots indicate distances for which the system was relaxed during the tip-lifting process. The numbers on the graph denote the number of layers in the capillary column (for the eight-layer column this number is ambiguous due to disordering in the middle part, see figure 8*c*).

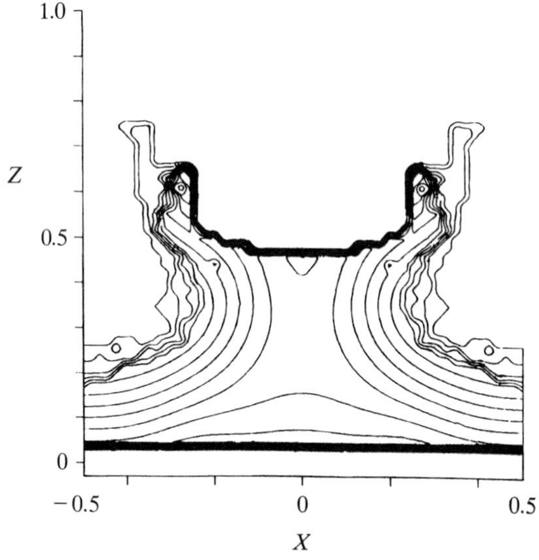

Figure 11. Segmental density contours for the 36.9 Å high capillary column (see figure 8*c*) of hexadecane molecules, obtained via MD simulations at 350 K. The density at the middle (core) of the column is 0.033 segments $Å^{-3}$, which is the same as that of bulk hexadecane at the above temperature. The spacing between contour lines $\Delta\rho = 0.003$ segments $Å^{-3}$. The linear dimension along the x and z directions is 77.5 Å.

liquid layer spacings. Upon raising the tip to a distance $d_{ts} = 36.9$ Å (figure 8*c*) and subsequently to $d_{ts} = 45.7$ Å (figure 8*d*) the layered nature of the confined film (in the middle region of the capillary column) diminishes. The short-time trajectories of hexadecane molecules at the tip, middle and bottom of the liquid column of height 36.9 Å, shown in figure 9, illustrate the ordered nature of the film next to the tip and surface, and reduced order at the middle of the capillary junction.

The transition from a layered to a partially layered liquid junction is reflected in the force against distance curve shown in figure 10, portraying the energetics of the system (compare with figure 6, corresponding to the elongation of solid junctions, where no such transition was observed). Clearly, the force required in order to elongate the column in the height range corresponding to layered configurations is significantly larger than that needed to separate the two interfaces when part of the liquid junction between them acquires liquid-like properties.

Finally, such simulations allow a quantitative analysis of the density distribution and energetics of the capillary structures on the molecular level, as illustrated in figure 11, where contours of the segmental density for a capillary column of height $d_{ts} = 36.9$ Å, are shown. First, it is of interest to note that even for such microscopic capillary junctions the density at the core achieves that of the bulk liquid at the same temperature (T = 350 K). Secondly, the density profile of the column exhibits a gradual decrease from the core outwards. The radii of curvature determined from such plots together with contours of the molecular pressure tensor distribution in the column, allow a direct microscopic evaluation of the Young–Laplace equation (Luedtke & Landman 1992) and disjoining pressure of the film, and provide detailed molecular-level guidance for the analysis of AFM data (Mate *et al.* 1989; Mate & Novotny 1991).

Research supported by U.S. Department of Energy, the Air Force Office of Scientific Research and the NSF. Calculations performed at the Florida State Computer Center and at NERSC, Livermore, California, through a grant by DOE.

References

Adams, J. B., Foiles, S. M. & Wolfer, W. G. 1989 Self-diffusion and impurity diffusion of fcc metals using the five-frequency model and the embedded atom method. *J. Mater. Res.* **4**, 102–112.

Barnett, R. N., Landman, U., Makov, G. & Nitzan, A. 1990 Theoretical studies of the spectroscopy of excess electrons in water clusters. *J. chem. Phys.* **93**, 6226–6238.

Barnett, R. N., Landman, U. & Rajagopal, G. 1991 Patterns and barriers for fission of charged small metal clusters. *Phys. Rev. Lett.* **67**, 3058–3061.

Behm, R. J., Garcia, N. & Rohrer, H. (eds) 1989 *Scanning tunneling microscopy and related methods.* Dordrecht: Kluwer.

Benedek, G., Martin, T. P. & Pacchioni, G. (eds) 1988 *Elemental and molecular clusters.* Berlin: Springer.

Bonacic-Koutecky, V., Fantucci, P. & Koutecky, J. 1991 Quantum chemistry of small clusters of elements of groups Ia, Ib and IIa: fundamental concepts, predictions and interpretation of experiment. *Chem. Rev.* **91**, 1035–1070.

Bowden, F. P. & Tabor, D. 1973 *Friction.* Garden City, New York: Anchor Press / Doubleday.

Burnham, N. A., Colton, R. J. & Pollock, H. M. 1991 Interpretation issues in force microscopy. *J. Vac. Sci. Technol.* A **9**, 2548–2556.

Catlow, C. R. A. & Mackrodt, W. C. (eds) 1982 *Computer simulations of solids.* Berlin: Springer.

Desaguliers, J. T. 1734 *A course of experimental philosophy*, vols 1 and 2. London.

Dieter, G. 1967 *Mechanical metallurgy.* New York: McGraw-Hill.

Dowson, D. 1979 *History of tribology.* London: Longman.

Israelachvili, J. N. 1992 *Intermolecular and surface forces* (2nd edn). London: Academic Press.

Jena, P., Rao, B. K. & Khanna, S. N. (eds) 1987 *Physics and chemistry of small clusters.* New York: Plenum.

Kaukonen, H.-P., Barnett, R. N. & Landman, U. 1992 Dielectrons in water clusters. *J. chem. Phys.* **97**, 1365–1377.

Landman, U. & Luedtke, W. D. 1991 Nanomechanics and dynamics of tip–substrate interactions. *J. Vac. Sci. Technol.* B **9**, 414–423.

Landman, U. & Luedtke, W. D. 1992 Consequences of tip–substrate interactions. In *Scanning microscopy III* (ed. H.-J. Guntherodt & R. Wiesendanger). Berlin: Springer.

Landau, D. P., Mon, K. K. & Schuttler, H.-B. (eds) 1988 *Computer simulation studies in condensed matter physics*. Berlin: Springer.

Landman, U., Scharf, D. & Jortner, J. 1985 Electron localization in alkali-halide clusters. *Phys. Rev. Lett.* **54**, 1860–1863.

Landman, U., Luedtke, W. D., Burnham, N. A. & Colton, J. 1990 Atomistic mechanisms and dynamics of adhesion, nanoindentation, and fracture. *Science, Wash.* **248**, 454–461.

Landman, U., Luedtke, W. D. & Ringer, E. M. 1992 Atomistic mechanisms of adhesive contact formation and interfacial processes. *Wear* **153**, 3–30.

Leggetter, S. & Tildesley, D. J. 1989 The computer simulations of adsorbed hydrocarbons. *Molec. Phys.* **68**, 519–546.

Luedtke, W. D. & Landman, U. 1992 Solid and liquid junctions. *Comp. Mater. Sci.* **1** (In the press.)

Mate, C. M., Lorenz, M. R. & Novotny, V. J. 1989 Atomic force microscopy of polymeric liquid films. *J. chem. Phys.* **90**, 7550–7555.

Mate, C. M. & Novotny, V. J. 1991 Molecular conformation and disjoining pressure of polymeric liquid films. *J. chem. Phys.* **94**, 8420–8427.

Murday, J. S. & Colton, R. J. 1990 Proximal probes: techniques for measuring at the nanometer scale. *Mater. Sci. Engng* B **6**, 77–85.

Nieminen, R. M., Puska, M. J. & Manninen, M. J. (eds) 1990 *Many-atom interactions in solids*. Berlin: Springer.

Pollock, H. M. & Singer, I. (eds) 1992 *Fundamentals of tribology*. Dordrecht: Kluwer.

Preston, M. A. & Bhaduri, R. K. 1975 *Structure of the nucleus*, p. 589. Reading, Massachusetts: Addison-Wesley.

Rajagopal, G., Barnett, R. N., Nitzan, A., Landman, U., Honea, E. C., Labastie, P., Homer, M. L. & Whettne, R. L. 1990 Optical spectra of localized excess electrons in alkali halide clusters. *Phys. Rev. Lett.* **64**, 2933–2936.

Rajagopal, G., Barnett, R. N. & Landman, U. 1991 Metallization of ionic clusters. *Phys. Rev. Lett.* **67**, 727–730.

Ribarsky, M. W. & Landman, U. 1992 Structure and dynamics of n-alkanes confined by solid surfaces I. Stationary crystalline boundaries. *J. chem. Phys.* **97**, 1937–1949.

Ryckaert, J. P. & Bellemans, A. 1978 Molecular dynamics of liquid alkanes. *Faraday Discuss. chem. Soc.* **66**, 95–106.

Sarid, D. 1991 *Scanning force microscopy*. New York: Oxford University Press.

Scharf, D., Jortner, J. & Landman, U. 1987 Cluster isomerization induced by electron attachment. *J. chem. Phys.* **87**, 2716–2723.

Vitek, V. & Srolovitz, D. J. (eds) 1988 *Atomic simulation of materials: beyond pair potentials*. New York: Plenum.

Xia, T. K., Ouyang, J., Ribarsky, M. W. & Landman, U. 1992 Interfacial alkane films. *Phys. Rev. Lett.* (In the press.)

Ab initio cluster calculations of defects in solids

By R. Jones

Department of Physics, University of Exeter, Exeter EX4 4QL, U.K.

A method based on local density functional theory is described which leads to the rapid determination of the structure, vibrational and electronic properties of clusters as large as 100–150 atoms. The technique is particularly suitable for molecular solids, covalently bonded materials where the clusters are terminated by hydrogen, and to ionic systems where the termination consists of a set of distributed charges. The strengths and weaknesses of the method are detailed together with an application to the interstitial carbon–oxygen complex in silicon where oxygen is found to be over-coordinated. The good agreement obtained for the vibrational modes of the complex lends support to the unusual structure found.

1. Introduction

The determination of the properties of defects in solids – and I have in mind complicated defects like dislocations and impurity aggregates – poses severe problems. First, one needs to be able to describe correctly the charge distribution around each atom and to evaluate the force acting on it. Thus a method of solving the many body Schrödinger equation is required. The two standard methods: Hartree–Fock (HF) and local density functional (LDF) theories (Lundqvist & March 1986; Ihm 1988) are not devoid of approximations and assumptions but they have been found to be particularly useful for ground state molecular and crystalline structures. Each is a variational procedure with HF theory assuming the wave-function, which is dependent on the coordinates of all the electrons, as the variational variable whereas LDF theory takes it to be the charge density: a function of just three coordinates (in the spin-polarized version, the variational variables include the magnetization density). Both theories can be written in terms of single particle Schrödinger equations with the potential acting on a electron arising from an effective field due to all the others. Thus both require a self-consistent equation to be solved. However, there are important differences; especially in the treatment of exchange and correlation. HF theory ignores the latter and its inclusion via say Möller–Plesset perturbation theory is unwieldy. For metallic systems correlation is essential and for this reason HF methods have primarily been used in insulators. LDF theory includes a correlation term derived from the homogeneous electron gas but its utility in multi-atomic systems where the charge density varies rapidly is well proven. The exchange energy in HF theory is a four-centre integral and its evaluation requires the computation of $O(N^4)$ integrals, where N is the basis size. This is to be contrasted with LDF theory whose exchange-correlation energy is an integral of a function of the electron density $n(\boldsymbol{r})$ and its evaluation requires $O(N^2M)$ computations where M is the number of points or operations involved in estimating this integral. In many applications this scales as N or the cluster size. We should say that this applies to

Phil. Trans. R. Soc. Lond. A (1992) **341**, 351–360
Printed in Great Britain
©1992 The Royal Society

clusters of about 100 atoms, where typically each basis function has some overlap with almost all the others. Thus we expect LDF theory to be about N times faster than HF and for typical values of N around 500 to 1000, this makes HF theory slower for large systems. Nevertheless, the latter theory has been used with some success for large systems (Maric *et al.* 1989; Nada *et al.* 1990).

A second reason for preferring LDF over HF theory is that for many purposes it is only the valence electrons that are important in chemical bonding. The development of pseudopotentials (Bachelet *et al.* 1982; Yin & Cohen 1982) eliminating the need to include core electrons has been successfully accomplished in the case of LDF theory. The total electron density is composed of two parts: a core density which is the sum of contributions from different atomic cores and is large near each atom but falls off rapidly to zero, and the valence charge density which although varying rapidly near the core (because of the constraints imposed by orthogonalization) is relatively smooth around the centres of chemical bonds. The pseudo-atoms have no core states and thus their 'valence' electrons experience a repulsive potential leading to a small and slowly varying charge density in the core region. In the frozen-core approximation, the exchange-correlation energy for the pseudo-atom is then determined by this valence charge density alone. This is of great significance for it is a difficult task to construct a basis set for both the core and valence wave-functions as each of these quantities has a different domain of importance and scale of variation. The result is that it is of no greater difficulty in treating say GeO_2 than SiO_2.

The nature of the exchange energy in HF theory involving the product of four orbital functions some of which may be core ones has made it more difficult to develop reliable pseudopotentials. Thus the most efficient way of treating defects in semiconductors may well be based on LDF theory incorporating pseudopotentials and thereby eliminating core electrons. However, some quantities dependent on the core wave-functions, such as chemical shifts, may not then be calculable. This is not always the case as Van de Walle (1990) finds good agreement for the relative hyperfine and super-hyperfine parameters (quantities depending on the wave-function near the nucleus) for H in Si.

Finally, one requires an efficient method of solving the Schrödinger equation and, in addition, it is essential to be able to calculate the forces acting on individual atoms and allow the positions of these to adjust until an equilibrium structure is found.

There appear to be three methods: first, the Green function method (Baraff & Schlüter 1983) which is the most rigorous one for point defects. Much effort is expended on the evaluation of the host lattice Green function, more elements of which are required when the size of the defect increases. Second, the supercell method which uses a basis of plane-waves together with the molecular dynamical method of Car & Parrinello (1984) and seems ideal for many problems. There are difficulties for certain elements, e.g. O, F and transition ones, due to the lack of p or d core electrons as this makes the valence wave-functions vary rapidly near the cores with the consequence that one requires an extremely large number of plane-waves. A second difficulty is that for certain defects, e.g. partial dislocations in semiconductors, it is necessary for topological reasons to construct unit cells containing dislocation dipoles. The present limitation of unit cell size causes the dislocations to lie unsatisfactorily close together. In other systems such as zeolites or proteins where the unit cells contain hundreds or even thousands of atoms it may not be easy to construct a sufficiently small cell for computational purposes. The third method is based on atomic clusters. The essential problem here is to passivate the surface of the cluster in such a way that properties of the inner part are insensitive to the termination. For non-metallic solids this requires a surface without gap states whose charging would draw charge from the inner part of the cluster. This would affect the properties of the inner atoms. A practical way of doing this is to passivate surface dangling bonds of covalently bonded materials with hydrogen. Moreover it is important to choose a short H-surface length as this depresses the H-bonding states below the bulk valence band top and elevates the H-antibonding states to above the bulk conduction band. However, since the host feels a repulsive potential from the surface H atoms, its valence and conduction bands are also depressed and elevated respectively resulting in an increased band gap. This effect of H in increasing the band gap is realized in a-Si:H and, possibly, porous Si. However, it seems that this band gap widening does not significantly affect structural or vibrational properties, although it does lead to defect levels lying deeper in the gap than observed. It is known that small H-terminated molecules have structures and vibratory modes close to those of the bulk. For example, neopentane C_5H_{12} has a bond length within 1% of diamond whereas disiloxane, $(SiH_3)_2O$, has an Si–O length of 0.1634 nm and vibratory modes at 1107 and 606 cm^{-1}, which lie close to those of interstitial oxygen in Si: 0.16 nm, 1136 and 515 cm^{-1} (Stavola 1984). The termination must be different for ionic systems and here a distribution of fixed charges has been used to surround the cluster.

I shall describe the cluster method that I have developed in more detail in the next section. Note here that several other workers have also developed cluster LDF methods (Pederson & Jackson 1990) but not incorporating pseudopotentials. I shall give what I see as the strengths and weaknesses of the method in §3 and an application in §4. Before this I make some remarks on approximate methods. These are like CNDO, MNDO, PRDDO or the tight binding scheme where approximations to the HF or LDF theories are made at the outset. Very often these include empirical information and, provided that the bonding is properly described, give these methods a wide domain of applicability and usefulness. In particular they can give stretch frequencies *systematically* higher than those observed and this is a very useful result.

2. The LDF cluster method

The total energy of the cluster is given by the minimum of

$$E = E_{\mathrm{ke}} + E_{\mathrm{e\text{-}p}} + E_{\mathrm{H}} + E_{\mathrm{xc}} + E_{\mathrm{i\text{-}i}}.$$

Here E_{ke}, $E_{\mathrm{e\text{-}p}}$, E_{H}, E_{xc}, and $E_{\mathrm{i\text{-}i}}$ are the kinetic, electron–pseudopotential, Hartree, exchange-correlation and ion–ion energies respectively. The charge density $n(\boldsymbol{r})$ is written in terms of the wave-functions, $\psi_\lambda(\boldsymbol{r})$, of occupied states, each of which is expanded in a basis set $\phi_i(\boldsymbol{r})$ of localized orbitals.

It is then necessary to vary the coefficients in the wave-functions in order to minimize E subject to the usual constraint that the total number of electrons is fixed. This is achieved by writing Euler–Lagrange equations for these coefficients. These equations can be cast in a matrix form involving integrals over the basis functions. We have, for the λ wave-function:

$$\psi_\lambda(\boldsymbol{r}) = \sum_i c_i^\lambda \phi_i(\boldsymbol{r}),$$

and the set of c_i^λ which minimize E, subject to the constraint that the cluster contains the correct number of electrons, satisfy (Jones & Sayyash 1986):

$$\sum_j (KE_{i,j} + V^p_{i,j} + V^H_{i,j} + \mu_{i,j} - E_\lambda S_{i,j})\, c_j^\lambda = 0.$$

Here $KE_{i,j}$, $V^p_{i,j}$, $V^H_{i,j}$, $\mu_{i,j}$, and S_{ij} are the matrix elements for the kinetic, pseudopotential, Hartree and exchange-correlation potentials and overlap respectively.

Gaussian basis functions have the advantage that all the required integrals can be analytically performed. It is trivial to evaluate the integrals involving the kinetic energy, and the pseudopotential; especially if one uses those given by Bachelet *et al.* (1982). This leaves the two troublesome terms: the Hartree energy, E_{H}, and the exchange-correlation energy, E_{xc} where

$$E_{\mathrm{H}} = \frac{1}{2}\int \frac{n(\boldsymbol{r}_1)\, n(\boldsymbol{r}_2)}{|\boldsymbol{r}_1 - \boldsymbol{r}_2|}\, \mathrm{d}\boldsymbol{r}_1\, \mathrm{d}\boldsymbol{r}_2$$

and

$$E_{\mathrm{xc}} = \int \epsilon_{\mathrm{xc}}(n)\, n\, \mathrm{d}\boldsymbol{r},$$

where ϵ_{xc} is the exchange-correlation energy density.

For large systems it is not possible to treat these terms exactly. Consequently some approximation must be used. It is important that the approximations used in evaluating E_{H} and E_{xc} are consistent with those in the Hartree and exchange-correlation potentials otherwise the self-consistent density would not be the one that minimizes the total energy.

To achieve this we replace E_{H} and E_{xc} with approximate expressions $\tilde{E}_{\mathrm{H}}$ and $\tilde{E}_{\mathrm{xc}}$ (Jones & Sayyash 1986; Jones 1988). Here

$$\tilde{E}_{\mathrm{H}} = \int \frac{n(\boldsymbol{r}_1)\, \tilde{n}(\boldsymbol{r}_2)}{|\boldsymbol{r}_1 - \boldsymbol{r}_2|}\, \mathrm{d}\boldsymbol{r}_1\, \mathrm{d}\boldsymbol{r}_2 - \frac{1}{2}\int \frac{\tilde{n}(\boldsymbol{r}_1)\, \tilde{n}(\boldsymbol{r}_2)}{|\boldsymbol{r}_1 - \boldsymbol{r}_2|}\, \mathrm{d}\boldsymbol{r}_1\, \mathrm{d}\boldsymbol{r}_2$$

and

$$\tilde{E}_{\mathrm{xc}} = \int \epsilon_{\mathrm{xc}}(\tilde{n})\, \tilde{n}\, \mathrm{d}\boldsymbol{r}.$$

It is clear that these expressions are exact when $\tilde{n} = n$.

We define the approximate density $\tilde{n}(\boldsymbol{r})$ in terms of basis functions $g_k(\boldsymbol{r})$ by

$$\tilde{n}(\boldsymbol{r}) = \sum_k c_k\, g_k(\boldsymbol{r}).$$

It is most sensible to choose the coefficients c_k by requiring the difference between $\tilde{E}_{\mathrm{H}}$ and E_{H} to be as small as possible (Dunlap *et al.* 1979). This difference can be written as

$$\frac{1}{2}\int \frac{(n(\boldsymbol{r}_1) - \tilde{n}(\boldsymbol{r}_1))(n(\boldsymbol{r}_2) - \tilde{n}(\boldsymbol{r}_2))}{|\boldsymbol{r}_1 - \boldsymbol{r}_2|}\, \mathrm{d}\boldsymbol{r}_1\, \mathrm{d}\boldsymbol{r}_2.$$

Then the coefficients c_k are chosen to minimize this expression. In practice we choose some g_k to be the functions $(1 - \frac{2}{3}a_k(\boldsymbol{r} - \boldsymbol{R}_k)^2)\exp(-a_k(\boldsymbol{r} - \boldsymbol{R}_k)^2)$ and others just s-gaussians centred at $\boldsymbol{R}_k$. The point about the first set is that they give a potential which is just $(2\pi/(3a_k))\exp(-a_k(\boldsymbol{r} - \boldsymbol{R}_k)^2)$ and thus the matrix elements $\tilde{V}^H_{i,j}$ are trivial to compute (Jones 1989). Since, however, each of the first set of functions integrates to zero it is essential to include some functions of the second type. The sites $\boldsymbol{R}_k$ are taken to lie at both atomic and bond centres.

To evaluate $\tilde{E}_{\mathrm{xc}}$ we use

$$\tilde{E}_{\mathrm{xc}} = \sum_k c_k \int g_k(\boldsymbol{r})\, \epsilon_{\mathrm{xc}}(\tilde{n})\, \mathrm{d}\boldsymbol{r}$$

and estimate the integrals using the first few moments of $\tilde{n}$ over g_k. We do not need to use the same coefficients c_k and functions g_k for $\tilde{n}$ and instead use a least squares fit to n using simple gaussian functions.

The key remark is then that the Euler–Lagrange equations for the wave-function are then derived from minimising this approximate energy expression. Thus the self-consistent density is necessarily the one giving the lowest energy $\tilde{E}$. This would not be the case if one used different approximations to evaluate say E_{H} and the matrix elements of the electrostatic potential V^H. Another advantage of the formulation in terms of analytic integrals is that the energy enjoys the full point group symmetry of the cluster which is not necessarily the case if the matrix elements were evaluated by a numerical integration over a cartesian mesh.

The above formulation allows the forces to be calculated semi-analytically using the Hellmann–Feynman theorem and numerical estimates of all the derivatives. It is important, however, to include the derivatives of the basis functions.

We have used basis sets of s and p gaussian orbitals although code has recently been written which extends these to any polynomial multiplied by a gaussian function. The second derivatives of the energy can also be calculated and used to obtain the vibrational modes of defects (Jones *et al.* 1991).

3. The strengths and weaknesses of the cluster method

For solid state problems there are a number of cases where cluster theory gives useful information but, in others, it is seriously unreliable.

For molecular solids, i.e. those composed of largely non-polar molecules like zero-dimensional phosphorous–selenide glasses (Jones & Lister 1989) or the fullerenes (Jones *et al.* 1992), the method is ideal as it can concentrate on the properties of the isolated molecules. This could also be true for polymers or liquid crystals. For covalently bonded solids or defects within them, it appears that H-termination is an extremely effective means of passivating the surface. The result is that structural and the higher phonon branches are well described. The lower modes and the elastic constants require longer ranged interactions (Kunc 1985) and so are more difficult to describe with the theory. The cohesive energy of the solid is also much more difficult to obtain because the basis can often be incomplete and because of the presence of the H atoms. The electronic gaps are, as stated above, much larger than they should be as it is well known that LDF theory leads to gaps smaller than those observed.

For point defects such as impurities or impurity aggregates, the method successfully gives structures in agreement with other LDF calculations (Jones 1989) but in addition the local modes of vibration of light impurities are reasonably well described. The migration or reorientation energy within the cluster can also be found such as for O in Si (Jones *et al.* 1991). The relative energies of H in various positions of the Si lattice agree well with other LDF calculations (Briddon & Jones 1990) but bond energies are usually too large. Recent work has focused on dislocations and their interaction with impurities (Heggie *et al.* 1992).

Graphite, a layered material, can also be treated with H atoms terminating layers of carbon atoms. Both interplanar and intraplanar lengths are determined to within a few percent of the observed values (C. D. Latham, personal communication).

Table 1. *Frequencies (in* cm^{-1}*) of local modes of* C_i–O_i

$^{12}C^{16}O^{28}Si$	$^{12}C^{18}O^{28}Si$	$^{13}C^{16}O^{28}Si$	$^{14}C^{16}O^{28}Si$	$^{12}C^{16}O^{30}Si$
calculated				
1141	1140	1101	1067	1136
925	925	898	874	917
625	624	624	624	604
604	598	604	604	589
565	562	564	564	550
559	558	558	558	541
observed				
1115.5	1115.5	1078.3	1047	
865.2	865.2	841.8	819.2	
742				
586 (72.6 meV)	582	586	586	576
550				
528 (65.5 meV)	523	528	528	

For H-bonded systems like ice, recent work (Heggie *et al.* 1992) indicates that the O–H distance between atoms belonging to different water molecules decreases with the cluster size as expected. This is an exciting discovery since it shows that the method may be applicable to biochemically important materials.

For silicates like quartz, the method gives excellent Si–O lengths and Si–O–Si, O–Si–O angles and the energy derivatives of distorted clusters can be used to fit classical potentials such as those of the Catlow–Sanders form. Thus has given very encouraging results for the properties of quartz under pressure (J. Purton, R. Jones, C. R. A. Catlow & M. Leslie, unpublished results). Ionic solids are best treated with a termination consisting of charge distributed around sites outside the cluster. For NaCl, MgO and Al_2O_3 the bond lengths are given to within few percent.

Thus there are a wide range of materials and their properties that can be explored using the cluster method.

The computationally most intensive routines are those involved in evaluating the three centred integrals but it is these which are most easy to vectorize and parallelize. Consequently, it is possible to relax an inner set of say 17 atoms of a 70 atom cluster in about three CPU days on an IBM RS6000 work station and evaluate the necessary second derivatives in about the same amount of time. This represents nowadays a modest computational requirement.

The ability to compute the forces on the atoms near the defect and to move them until equilibrium prevails has given in several cases quite unexpected results. It is these cases that are of the greatest interest for they show that our intuition has led us astray and this might then suggest a resolution of some long standing problem. I next describe an example of this sort which has given quite unexpected results.

4. Application to the interstitial carbon–oxygen defect in silicon

This is a complicated but important defect as it is one of the dominant defects produced in electron irradiated Si which contains C and O impurities. The defect is known to contain one O and C atom and its vibrational modes (Davies *et al.* 1986) are given in table 1. What is surprising is that the highest mode at 1115 cm^{-1} is unaffected by ^{18}O doping quite unlike bond centred interstitial oxygen (Newman 1973).

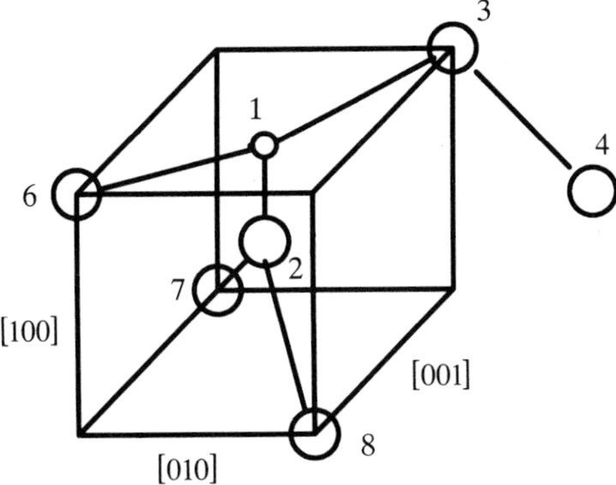

Figure 1. Schematic arrangement of atoms in the C_i split-interstitial. The small circle denotes a C atom. Note the dangling bonds on C and the central Si atom.

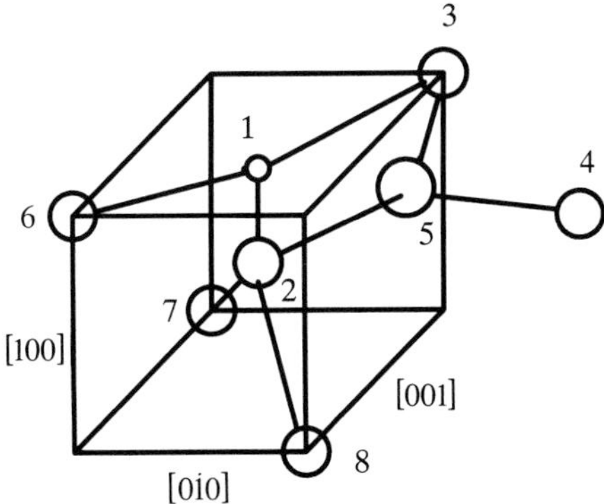

Figure 2. Schematic arrangement of atoms in C_i–O_i complex. The small circle represents the C atom and the large one the over-coordinated O atom.

Magnetic resonance experiments (Trombetta & Watkins 1987) showed that the structure of the defect must be closely related to that of the C split-interstitial (Watkins & Brower 1976) illustrated in figure 1, and they suggested that the C atom has a filled dangling bond parallel to, say $[01\bar{1}]$, and is bonded to a Si atom with an empty dangling bond nearly parallel to [011], with O occupying a bond centred location between atoms 3 and 4 of figure 1. Now this model would assign the 1115 cm^{-1} mode to an O stretch mode which is inconsistent with the insensitivity of this mode to ^{18}O substitution.

The LDF cluster calculations (Jones & Öberg 1992) revealed very unusual bonding in the defect. The C, O and 17 Si atoms of a 73 atom H-terminated cluster $COSi_{35}H_{36}$ centred on the Trombetta–Watkins model of the defect were relaxed. We found that the O-atom moved away from the bond centred position towards the middle of the quadrilateral (in the $(01\bar{1})$ plane) formed by atoms 1, 2, 3 and 4 in figure 2. The reason is that the Si dangling bond has an empty level lying above that of C. It can lower the energy of one set of lone pair electrons on O (atom 5) by forming a dative bond and pulling O towards it. Consequently O becomes over-coordinated. This model explains why O prefers to bind to the defect in the same plane as the Si dangling bond.

The vibrational modes of O are expected to be low lying because of the very long

Si–O lengths (*ca.* 0.185 nm) present in the over-coordinated O defect. The energy second derivatives between the atoms 1 to 8 in figure 2 were calculated and inserted into the dynamical matrix. The calculated frequencies of the six highest modes of C_i–O_i and their isotopic shifts are given in table 1.

The absolute values of the frequencies are within about 100 cm^{-1} of the observed ones but their isotopic shifts are given very well. This strongly suggests that the structure found here is the correct one.

To conclude, the calculations showed that interstitial O is unstable when close to a Si dangling bond and readily forms an over-coordinated defect with rather long Si–O bonds and low lying Si–O stretch modes. This is of interest to the more complex problem of O precipitation where Si interstitials might rise from the precipitation process and interact with other O complexes. Presumably some weak or broken bonds will be formed in these processes. These may well be attacked by O and form over-coordinated defects with low lying vibrational modes. I point out that no high frequency O related modes have been found for the thermal donor and it is tempting to speculate that this may be because the O atoms in that case are also over coordinated. There have been models of thermal donors along these lines (Chadi 1990; Jones 1990; Deak *et al.* 1991).

5. Conclusions

I have shown in this paper that the cluster LDF method can reveal the structure and electronic properties of a wide range of solids and especially their defects. The calculations have in several cases given surprising and unexpected results which may – and often does – explain some interesting phenomena. I believe that the method can be profitably used for a variety of defects and defect processes in many materials and enjoys a considerable advantage, for a first-principles one, in that it requires only a moderately significant computational effort.

It is a pleasure to acknowledge helpful discussions with a large number of colleagues. Special thanks are owed to P. R. Briddon, M. I. Heggie, G. M. S. Lister, C. D. Latham, S. Maynard, R. C. Newman, S. Öberg, M. Stoneham, V. Torres and A. Umerski.

References

Bachelet, G. B., Hamann, D. R. & Schlüter, M. 1982 Pseudopotentials that work: from H to Pu. *Phys. Rev.* B **26**, 4199–4228.

Baraff, G. A. & Schlüter, M. 1983 Total energy of isolated point defects in solids. *Phys. Rev.* B **28**, 2296–2298.

Briddon, P. R. & Jones, R. 1990 Hydrogen and muonium in silicon. *Hyperfine Interactions* **64**, 593–602.

Car, R. & Parrinello, M. 1984 Unified approach for molecular dynamics and density functional theory. *Phys. Rev. Lett.* **55**, 2471.

Chadi, D. J. 1990 Oxygen–oxygen complexes and thermal donors in Si. *Phys. Rev.* B **41**, 10595–10603.

Davies, G., Oates, A. S., Newman, R. C., Woolley, R., Lightowlers, E. C., Binns, M. J. & Wilkes, J. G. 1986 Carbon-related radiation centres in Czochralski silicon. *J. Phys.* C **19**, 841–855.

Deak, P., Snyder, L. C. & Corbett, J. W. 1991 Silicon-interstitial-oxygen-interstitial complex as a model of the 450 °C oxygen thermal donor in silicon. *Phys. Rev. Lett.* **66**, 747–749.

Dunlap, B. I., Connolly, I. W. D. & Sabin, J. R. 1979 On some approximations in application of X-alpha theory. *J. chem. Phys.* **71**, 3396–3402.

Heggie, M. I., Jones, R. & Umerski, A. 1991 Interaction of impurities with dislocations in silicon. *Phil. Mag.* A **63**, 571–584.

Heggie, M. I., Maynard, S. C. P. & Jones, R. 1992 Computer modelling of dislocation glide in ice Ih. In *Proc. Physics and Chemistry of Snow and Ice* (ed. N. Maeno & T. Hondoh). Hokkaido University Press.

Ihm, J. 1988 Total energy calculations in solid state physics. *Rep. Progr. Phys.* **51**, 105–142.

Jones, R. 1988 The phonon spectrum in diamond derived from ab-initio local density functional calculations on atomic clusters. *J. Phys.* C **21**, 5735–5745.

Jones, R. 1989 *Ab initio* calculations of the structure and properties of large atomic clusters. *Molec. Simulat.* **14**, 113–120.

Jones, R. 1990 *Ab initio* calculations on thermal donors in Si: an over-coordinated O atom model for the NL10 and NL8 centres. *Semicond. Sci. Technol.* **5**, 255–290.

Jones, R. & Lister, G. M. S. 1989 *Ab initio* calculations on the structure and vibrational properties of some phosphorus–selenium molecules: applications to zero-dimensional glasses. *J. Phys.* A **1**, 6039–6048.

Jones, R., Öberg, S. & Umerski, A. 1991 Interaction of hydrogen with impurities in semiconductors. *Mater. Sci. Forum* **83–7**, 551–562.

Jones, R. & Öberg, S. 1992 Oxygen frustration and the interstitial carbon–oxygen complex in silicon. *Phys. Rev. Lett.* **68**, 86–88.

Jones, R., Latham, C. D., Heggie, M. I., Torres, V. J. B., Öberg, S. & Estreicher, S. K. 1992 *Ab initio* calculations of the structure and dynamics of C_{60} and C_{60}^{3-}. *Phil. Mag.* (In the press.)

Jones, R. & Sayyash, A. 1986 Approximations in local density functional calculations for molecules and clusters: application to C_2 and H_2O. *J. Phys.* C **19**, L653–L657.

Kunc, K. 1985 *Ab initio* calculations of phonon spectra in solids. In *Proc. Nato Advanced Summer Study Inst.* (*Electronic structure: dynamics and quantum structural properties of condensed matter*) series B, vol. 121 (ed. J. T. Devreese & P. Van Camp). Plenum: New York.

Lundqvist, S. & March, N. H. 1986 *Theory of the inhomogeneous electron gas.* London: Plenum Press.

Maric, Dj. M., Vogel, S., Meier, P. F. & Estreicher, S. K. 1989 Equilibrium configuration of bond centred H^{o} in GaAs. *Phys. Rev.* B **40**, 8545–8547.

Nada, R., Catlow, C. R. A., Dovesi, R. & Pisani, C. 1990 An *ab initio* Hartree Fock study of α-quartz and stishovite. *Phys. Chem. Minerals* **17**, 353–373.

Newman, R. C. 1973 *Infra-red studies of crystal defects.* London: Taylor and Francis.

Pederson, M. R. & Jackson, K. A. 1990 Variational mesh in quantum mechanical simulation. *Phys. Rev.* B **41**, 7453–7461.

Stavola, M. 1984 Infra-red spectra of interstitial oxygen in silicon. *Appl. Phys. Lett.* **44**, 514–516.

Trombetta, J. M. & Watkins, G. D. 1987 Identification of the interstitial carbon-interstitial oxygen-complex in Si. *Appl. Phys. Lett.* **51**, 1103–1105.

Van de Walle, C. G. 1990 Structural identification of hydrogen and muonium centers in silicon: first-principles calculations of hyperfine parameters. *Phys. Rev. Lett.* **64**, 669–672.

Watkins, G. D. & Brower, K. L. 1976 EPR observation of the isolated interstitial carbon atom in Si. *Phys. Rev. Lett.* **36**, 1329–1332.

Yin, M. T. & Cohen, M. L. 1982 Theory of *ab initio* pseudopotential calculations. *Phys. Rev.* **25**, 7403–7412.

Discussion

P. C. H. Mitchell (*Department of Chemistry, The University, Whiteknights, Reading RG6 2AD, U.K.*): An impressive feature of your *ab initio* calculations is that they reveal the probable presence of three-coordinate oxygen at a defect site. There are a number of classic neutral molecular complexes having over-coordinate oxygen, e.g. basic beryllium and basic zinc acetate; $M_4O(O_2C.CH_3)_6$, (M = Be,Zn). In these structures a four-coordinate oxygen atom, at the centre of a tetrahedron of four metal atoms, forms a bond to each metal atom. The two oxygen atoms of each acetate bind to two different zinc atoms; the acetates thereby act as bridging groups along each edge of the tetrahedron of metal atoms. These structures would seem to offer an excellent means of testing out the computational procedure with well-characterized neutral molecules.

R. JONES: This suggestion is a good one and one I would like to follow-up.

Hardware and quantum mechanical calculations

By E. Wimmer

BIOSYM Technologies Inc., 9685 Scranton Road, San Diego, California 92121-3752, U.S.A.

The remarkable progress in the architecture, speed and capacity of computer hardware continues to drive the development of quantum mechanical methods, thus allowing calculations on increasingly complex systems. Using high-end computers, accurate quantum mechanical all-electron studies are now possible for solids such as transition metal compounds containing about fifty atoms per unit cell. Pseudopotential plane-wave methods are being applied to unit cells with 400 silicon atoms, and organic molecules consisting of over 100 atoms have become tractable using *ab initio* methods. Smaller, yet still useful calculations can be carried out on workstations. The combination of graphics workstations and high-performance supercomputers, integrated in tightly coupled heterogeneous networks, has allowed the design of software systems with unprecedented convenience and visualization capabilities. Despite this progress, however, there is still an urgent need for new quantum mechanical methods which converge systematically to the exact solution of Schrödinger's equation while maintaining a reasonable scaling of the computational effort with the system size.

1. Introduction and historical perspective

During the past four decades the development of quantum mechanical methods for accurate calculations on solids and their surfaces has been strongly coupled to the progress in computer hardware. In fact, a variety of hardware aspects play a decisive role in this evolution, especially computer architectures, processor speed, memory size, and external storage devices. While operating systems, compilers, mathematical libraries, software analysis and optimization tools, and networking software represent other important aspects of a computing system, the focus of this contribution is on the interplay between hardware and theoretical methods.

To gain a perspective, the major milestones in the development of computer hardware are reviewed in the context of the development of computational solid state physics. In §2, the three dominant computer architectures are discussed, i.e. vector supercomputers, RISC workstations, and massively parallel machines. Examples of solid state as well as molecular cluster calculations illustrate the current capabilities. Additional remarks are then made on the role of three-dimensional graphics workstations. A section of possible future developments concludes this paper.

The first generation of viable computing machines for quantum mechanical calculations were vacuum-based electronic computers such as M.I.T.s Whirlwind I in the 1950s. The major application of this machine was fluid dynamics. Using spare computing time during the night, these machines were used for pioneering quantum mechanical calculations. The processor speed of this machine, which occupied two

Phil. Trans. R. Soc. Lond. A (1992) **341**, 361–371
Printed in Great Britain
© 1992 The Royal Society

floors of a building, was 0.02 MFlops and the memory size was initially 25000 16-bit words (Ralston & Meek 1976).

Although the invention of the transistor in 1948 by Bardeen, Brattain and Shockley initiated a major technological breakthrough for computer hardware, it was not until 1959 that the first transistor-based or second generation computers, such as the UNIVAC M460, the CDC 1604 and the IBM 7030, became available. With the introduction of the IBM 360 series in April 1964 and the CDC 6600 in the autumn of that year, two highly successful series of 'third generation' computers became more widely available to the growing community of computational solid state physicists. These third generation computers are characterized by the use of large scale integration (LSI) techniques for their semiconducting logic devices. Measured by today's standards, the capabilities of third generation computers were limited not only by slow processors, but also by their small and slow direct-access memories made of arrays of magnetic cores. For example, early versions in the IBM 360 series had only 128 kbytes of magnetic core memory, which is surpassed by more than an order of magnitude in today's lap-top computers. Nevertheless, this third generation of computer hardware, available in the late 1960s and into the 1970s enabled first-principles electronic structure calculations for compounds such as TiC in a sodium chloride structure containing two atoms per unit cell (see, for example, Neckel *et al.* 1976). These calculations provided a detailed understanding of the bonding mechanism and electronic charge distribution in these materials, allowing interpretations and predictions of properties such as X-ray absorption and photoemission spectra. In these early computations, full exploitation of symmetry was essential to keep the size of the hamiltonian matrices under the limit of 40×40.

Initially, these calculations were not self-consistent, solving the bandstructure problem for selected high-symmetry k-points for a given crystal structure and a given crystal potential, which typically was constructed from a superposition of atomic densities. In the late 1960s and throughout the 1970s, machines of increasing speed such as the Control Data CDC 7600 computers as well as further developments of the IBM 360 and 370 mainframes provided the hardware platforms for a large number of electronic structure calculations that were carried out self-consistently, but using a simplified shape for the electron density and potential in the form of the muffin-tin potential (Moruzzi *et al.* 1978). Pseudopotential theory evolved as a particularly successful approach for semiconductors (as reviewed by Cohen & Chelikowsky 1989) and a wealth of calculated data started to provide a systematic understanding of the electronic structures of bulk metals, semiconductors, and simple compounds.

Further progress was only partly due to improved hardware. Advances in theoretical methods and more efficient algorithmic implementations played an equally important role. For example, in the original augmented plane wave (APW) method (Slater 1937) the hamiltonian matrix elements depend on the energies. Consequently, the resulting eigenvalue problem is nonlinear and has to be solved by a tedious, discrete energy search. Similar computational complications occur in the Korringa–Kohn–Rostoker (KKR) method (Korringa 1947; Kohn & Rostoker 1954). The linearization of the APW and the KKR methods (Andersen 1975) constituted a significant step forward. Using a linearized approach, all eigenvalues and eigenvectors for a given k-point can be evaluated in one diagonalization step.

The computational effort in such a linearized method is spread between the evaluation of the hamiltonian and overlap matrix elements and the diagonalization.

The size of these matrices is linearly proportional to the number of atoms, N. The diagonalization involves a step that scales like N^3. Memory requirements scale like N^2. While computing time presents a soft limit, the available memory size poses a hard constraint on the maximum number of atoms per unit cell, unless one finds 'out-of-core' solutions. For example, the memory of a CDC 6600-class machine could hold about 130000 floating point numbers, which allowed storage of two matrices (the hamiltonian matrix and the overlap matrix) of the size of approximately 200×200 each. Assuming that about 50 LAPW basis functions are needed per atom, only four inequivalent atoms per unit cell could be treated, if no symmetry was taken into account. The smaller number of basis functions needed in the KKR and linearized muffin-tin-orbital (LMTO) methods presented some advantages in this respect.

The unprecedented speed of the CRAY-1 vector supercomputer, which was introduced in 1976, represents another milestone for quantum mechanical calculations. While the CDC Star-100 preceded the CRAY-1 as vector processor, the success of the CRAY-1 was largely due to its balance between three aspects: high scalar performance, fast memory access, and vector registers. The CRAY-1 architecture is based on a fairly small instruction set and thus can be seen as 'vector RISC' (reduced instruction set computing) machine. Compared with previous mainframes, the CRAY-1 offered a fairly large memory of 8 Mbytes. With this new generation of computer hardware came three major advances in all-electron calculations: (i) surface calculations became possible through slab geometries (Krakauer *et al.* 1978); (ii) the muffin-tin approximation to the shape of the charge density and the potential was removed through the introduction of 'full-potential methods' (Wimmer *et al.* 1981); and (iii) total energy calculations became possible for challenging systems such as transition metal surfaces (Weinert *et al.* 1982). Together, these new capabilities opened the door to first-principles investigations of important phenomena such as adsorption (Wimmer *et al.* 1982), surface magnetism (Freeman *et al.* 1982) and surface reconstructions (Fu *et al.* 1984).

The success of the large-scale vector supercomputers encouraged several hardware manufacturers such as Control Data Corporation (through its subsidiary ETA Systems) and the Japanese manufacturers Fujitsu, NEC, and Hitachi to pursue the development of vector supercomputers. IBM offered vector facilities as addition to its 3090 mainframe series. Besides these mainframe machines, vector mini-supercomputers such as Convex and Alliant computers found rapid acceptance in many research departments in the second half of the 1980s.

2. Current hardware trends and capabilities

At present, three major computer architectures determine the hardware for quantum mechanical calculations: (i) shared memory vector machines with multiple processors; (ii) RISC workstations with high scalar speed; and (iii) massively parallel machines with distributed memory. An additional aspect is the availability of three-dimensional graphics capabilities which are integrated in workstations.

Shared memory vector machines with multiple processors such as the CRAY Y-MP, the Fujitsu VP series, the NEX SX series, and new generations of IBM mainframes are continuations of the mainframe developments of the past four decades. At present, NECs 4-processor SX-3 is the fastest shared-memory supercomputer, delivering 90% of its peak 25.6 GFlops for the LINPACK benchmark, whereas the 16-processor CRAY Y-MP/C90 with a theoretical peak

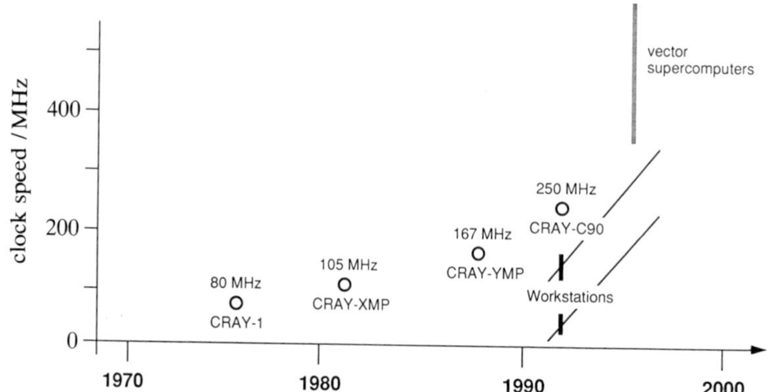

Figure 1. Development of processor speed shown for the example of CRAY supercomputers and typical workstations. Clock speeds are given for the first computer models in a series. Typically, the clock speed is increased during the lifetime of a series. The bar drawn for vector supercomputers around 1995 is an extrapolation based on current plans. Note that at present the processor speed of both supercomputers and workstations appears to increase at the same rate.

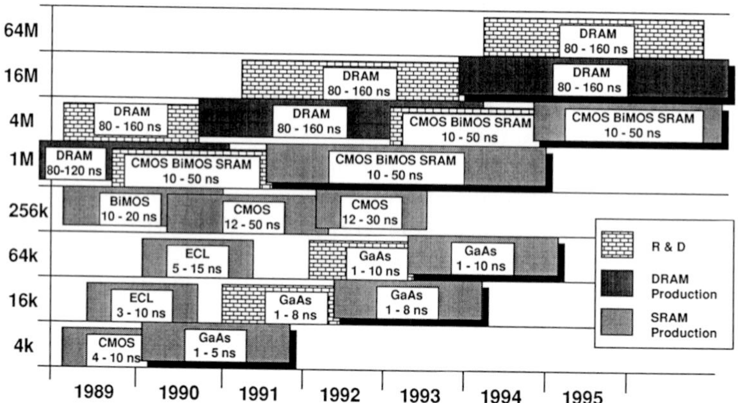

Figure 2. Development of size and access speed of memory components. SRAM stands for static random access memory and DRAM for dynamic random access memory. ECL refers to emitter coupled logic and CMOS is the abbreviation for complementary metal-oxide semiconductors. Note the trade-off between level of integration and access speed. (Courtesy of Cray Research, Inc.)

performance of 16 GFlops provides the greatest throughput for supercomputer work loads (Bell 1992). The maturity of the shared-memory architecture, the extremely high performance, and the familiarity with the programming model for these supercomputers makes these machines the work horses for large-scale quantum mechanical calculations as well as other applications involving solutions of linear algebraic equations.

RISC workstations such as the IBM RS/6000 are offering computing speed for individual researchers which is comparable with the performance of vector supercomputers a decade ago (such as the CRAY X-MP). While there is an impressive gain in speed and performance of workstations, the trend on the high-end supercomputers does not seem to be so different (cf. figure 1). Furthermore, increase in processor speed leads to higher system performance only if the memory access time can keep up. Figure 2 shows the trends in memory component technology. It can be seen that the development of the slower, but larger DRAM technology follows the same trend as the faster, but smaller CMOS SRAM components. Thus, the issues for machine performance are perhaps better understood if one considers computers not only from the standpoint of processors, but from the viewpoint of memory technology. A critical trade-off exists between large/small and slow/fast memory. Large and slow memories can be combined with slower processors, whereas fast memory is required by faster processors, but then the memory cannot be made so big. Thus, until recently, CRAY supercomputers had a relatively small, but very fast memory using emitter coupled logic (ECL) memory technology (cf. figure 2). For these reasons, the first CRAY Y-MP/8 machines had only 32 Mword of memory. Because of the improvements in the speed of CMOS memory components (cf. figure 2), today's CRAY supercomputers rely on larger CMOS memory technology. The CRAY-2 used a very large DRAM (and later SRAM) memory with only one memory port to the processors. Thus, the performance of the CRAY-2 is best when memory references are infrequent compared with the rate of floating point operations, but is relatively slow in the opposite case. Typical quantum mechanical solid state calculations represent the first case and thus the CRAY-2 continues to be a good, albeit somewhat unique architecture.

The actual improvements in performance can be seen from the following example. The CRAY Y-MP/C90, which was introduced in 1991, has a theoretical peak performance of 16 GFlops. Compared with the CRAY-1, which offered 160 MFlops, this represents an increase of computing speed of two orders of magnitude within 16 years. Significantly, during the same time, the memory size has increased by almost three orders of magnitude. Because of the maturity of this architecture and the existing experience with its programming model, a sustained performance of 10 GFlops for quantum mechanical calculations is now achievable on a machine such as the CRAY Y-MP/C90 without major changes of existing FORTRAN programs. Assuming a scaling of the computational effort with a third power as a function of system size, today's supercomputers would allow calculations for systems about five times as big as ten years ago. Alternatively, one could carry out calculations for 100 different geometries to map the energy hypersurface of a system where electronic structure calculations for only a single geometry were possible ten years ago. The actual progress is even larger because of significant improvements in computational methods and algorithms.

An example of current capabilities of quantum mechanical calculations is the determination of the electronic structure of an organometallic complex containing over one hundred atoms and a total of 1031 basis functions (Mertz & Andzelm 1991). Using a gaussian-based density function approach (Andzelm & Wimmer 1992) on an 8-processor CRAY Y-MP, a total speedup of 7.4 over a single processor was achieved. Table 1 provides an overview of the level of parallelization of the various steps in the calculation. It can be seen that already eight processors lead to a significant deviation from the ideal parallelization. Contrary to intuition, the analytic part of integral generation (three-centre integrals over gaussian functions) has the highest level of parallelization whereas the numerical grid generation shows relatively the lowest speedup. This may indicate that grid based methods are not necessarily 'naturally parallel'.

Workstations such as Apollo, SUN, Hewlett-Packard, and the IBM RS/6000 series introduced a new level of impressive performance at low cost. The key for this technology is a very high level of integration on the processor chip, combined with large and economic memory chips. Current IBM RS/6000 processors operate at a

Table 1. *Speedup due to parallelization of a density functional gaussian-type-orbital calculation of an organometallic Zn complex containing 112 atoms (1031 basis functions)*

(The results were obtained on an 8-processor CRAY Y-MP supercomputer using the DGauss program. The total CPU time to reach SCF convergence was approximately 40 minutes. XC fit refers to the numerical fitting to the exchange-correlation potential.)

task	processors 2	4	8
integrals	2	3.9	7.7
density fit	2	3.8	7.0
grid generation	2	3.7	6.2
XC fit	2	3.9	7.4
total	2	3.8	7.1

rate of about 40 MHz. The emerging generation of RISC processors such as the Digital Equipment Alpha chip will operate at about 150 MHz, thus surpassing the clock speed of the original CRAY-1 by almost a factor of two. As mentioned above, the speed of processors is just one of many hardware aspects. In fact, the development of RISC processors has shifted the bottleneck in the design of computer hardware even more from the processor to the issues of memory speed and communication (cf. figure 2).

The capabilities of workstations for quantum mechanical calculations are remarkable. For example, the self-consistent evaluation of the electronic structure for a molecule containing 52 atoms (560 basis functions) requires approximately 10 h of wall-clock time on a cluster of eight IBM RS/6000-320 workstations (S. Brode, personal communication). In these calculations, the workstations are coupled to form a parallel system. Compared with a single processor, a speed-up by a factor of 6.7 was achieved using an eight-processor cluster. In a similar calculation on an organometallic complex with 1017 basis functions, a speed-up of a factor of 10 was observed using a cluster of 14 processors (S. Brode, personal communication).

Another example of the computational capabilities of workstations is provided by a quantum mechanical calculation on a organometallic titanium complex modelling a catalytic reaction (M. Wrinn, unpublished results). The system, which has C_2 symmetry, contains 57 atoms and is described by 502 basis functions. Using a Silicon Graphics Personal Iris 35, the electronic structure and total energy has been evaluated with a density functional approach using numerical basis functions as implemented in the DMol program (Delley 1990) within about 10 h.

Massively parallel machines have captured the minds of many researchers in the field of computational quantum mechanics as well as other scientific disciplines, and there is no doubt that this architecture will play an increasingly important role in computational solid state physics and computational chemistry. The idea of parallel machines is by no means new. For example, the ILLIAC IV was built in 1967 with 64 processors. The design goal was 1 GFlop with 256 processors, but actually only 4 Mflop were achieved with 64 processors (Almasi & Gottlieb 1989). The reason was the use of pre-VLSI technology causing manufacturing and reliability problems. Another parallel architecture was implemented in the ICL DAP as described by Hockney & Jesshope (1981). Interestingly, before starting Cray Research, Seymour Cray worked on a shared memory scalar parallel machine as a successor for the CDC 7600. This parallel machine, which had a round shape and was internally called CDC

8600, was never completed. Instead, Seymour Cray designed the CRAY-1 with one processor, but vector registers.

Today's generation of parallel machines are based on VLSI technology and RISC processors. Currently there are a number of companies competing in this promising field including Intel, Thinking Machines, nCUBE, Kendall-Square, MasPar, Cray Research, IBM, and Meiko. It is now generally recognized that the key issue is not just processor speed, but rather the communication between various processing elements. The single instruction multiple data (SIMD) architecture turns out to be optimal only for specific applications whereas the multiple instruction multiple data (MIMD) architecture is more flexible and thus is emerging as the preferred approach. Memory topology and access speed across processing elements will be the decisive factors for success, as will be the convenience of the programming model and the corresponding system software.

Compared with today's shared memory multiprocessor vector machines and the RISC workstations, the architecture and usability of massively parallel machines is in an early stage. Nevertheless, despite the communication bottleneck, impressive results have already been obtained. The communication bottleneck can be seen in the following example. In a recent calculation of the 7×7 reconstruction of the Si(111) surface, Brommer *et al.* (1992) used a CM-2 parallel machine. In their calculations 25% of the total time was spent in a routine which sums numbers from all processors, in other words, gathering single numbers from all processors and adding them up. Independently, Stich *et al.* (1992) performed a first-principles structural and energetic determination using a similar approach using an Intel iPSC/860 hypercube and a 64-node Meiko i860 Computing Surface. In their work, both Brommer *et al.* and Stick *et al.* could explain the structure and energetics of this technologically important surface. Clearly, both calculations represent major milestones in electronic structure calculations.

3. Heterogeneous networks

In the previous section, three different architectures were discussed, shared memory parallel vector supercomputers, RISC workstations, and massively parallel machines. There are two other hardware aspects that need to be considered in the context of quantum mechanical calculations: the availability of three-dimensional (3D) graphics workstations and high-speed computer networks. The aim of the quantum mechanical calculations is not only a quantitative prediction, expressed in a few numbers such as a binding energy, lattice constants, and bulk modulus, but also physical insight into the atomic-scale phenomena of matter. To this end, visualization has become a significant component. In the early 1980s, graphics displays were often done on separate hardware devices such as the Evans & Sutherland Picture Systems and the transfer of data from the computational hardware to the graphics device was often cumbersome and slow. The introduction of 3D graphics workstations in the early to mid 1980s such as the Silicon Graphics, Stellar, and Ardent computers provided a new hardware concept by offering both impressive computational speed as well as convenient 3D-graphics capabilities. Compared with molecular biology, the use of 3D graphics and animation in computational condensed matter physics and chemistry is not yet fully developed and exploited, although impressive results have already been shown such as oxygen diffusion in bulk silicon (Joannopoulos 1992).

3D graphical user interfaces also allow a unique and unprecedented way to interact with the set-up and the execution of calculations. Most of the quantum mechanical calculations performed so far for solids and surfaces were done in batch jobs using text input files to direct the calculation. This is a tedious process which is quite susceptible to mistakes. Thus, large amounts of precious time is spent in tasks which are quite unrelated to the scientific investigation. This issue of scientific productivity is becoming more urgent as more large-scale numerical simulations are carried out in industrial research laboratories.

These factors have stimulated the development of integrated software systems combining the convenience of graphical user interfaces with the speed and computational capabilities of supercomputers. One such system, called UniChem, was developed at Cray Research with the aim to provide such a fully integrated quantum mechanical environment for molecular and cluster calculations. In this system, it is possible to build a molecular structure interactively on the workstation screen, then select a quantum mechanical method such as semi-empirical, density functional theory, or Hartree–Fock theory. The selection of the computational parameters is done in easily understandable pull-down menus and dialogue boxes. Launching a calculation across the network is literally one click on the mouse button. The progress of the calculation, for example a geometry optimization, can then be monitored on the workstation. Upon completion of the job on the supercomputer, the results can be immediately visualized including molecular structures, molecular orbitals, electron densities, spin densities, and electrostatic potentials. Normal vibrational modes can be visualized through animation of the molecule by selecting lines in the calculated vibrational spectrum.

While this system increases the productivity of expert users, it also allows – for better or worse – the non-expert to begin using these tools. From a user's point of view, the entire system behaves as a single entity, yet a number of different computer architectures are actually involved including the special-purpose graphics processors of the workstation, the vector capabilities of the shared-memory supercomputer, and the scalar processing power of the workstation for post-processing. Perhaps this type of transparently integrated system points into a direction of more generally integrated heterogeneous systems combining the three types of architectures discussed in the previous section.

4. Future

To predict future developments, it may be useful to state again the goals of quantum mechanical calculations of solids and surfaces. The aim could be described as the qualitative understanding and quantitative prediction of atomic-scale phenomena including geometric structures to within about ± 0.001 Å†, energy changes to within ± 0.01 eV for structural changes and ± 0.00001 eV for magnetic changes, the description of dynamic behaviour including chemical reactions with the ability to derive thermodynamic quantities such as free energies, and responses to external forces and electromagnetic fields. This should be possible for ordered as well as disordered systems. Full quantum mechanical descriptions of about 1000 atoms per system or unit cell are meaningful. Beyond that, it should be possible to embed the domain of the quantum mechanical description into the effective field created by the environment, where the environment could be treated by a force-field approach or even a continuum model.

† 1 Å = 10^{-10} m = 10^{-1} nm.

While current methods and hardware clearly do not meet the above goals, there is reason for optimism. For example, ground state structures such as lattice constants can be predicted to within about ± 0.02 Å, systems containing 400 atoms have been treated, and relative energies of the Si(111) 3×3, 5×5, and 7×7 reconstructed surfaces have been predicted from first principles calculations with energy differences of only about 0.02 eV per surface atom (Stich *et al.* 1992; Brommer *et al.* 1992). Perhaps the biggest challenges of quantum mechanical calculation of atomic assemblies is the formulation of theoretical methods that allow systematic convergence to the exact many-body result while maintaining a reasonable scaling of the computational effort. Traditional quantum chemical methods fulfil the first criterion, but not the second. Density functional fulfils rather the second than the first criterion. Definitely, more theoretical investigations are urgently needed. In the exploration of new ideas, the hardware has to be flexible, easy to use, and fast enough to enable the testing of methods and approaches.

Since the pioneering work of Slater (1951), the use of concepts from the homogeneous interacting electron gas as formalized in density functional theory has been the predominant many-body approach in quantum mechanical calculations for solids. The new implementation of gradient corrections to the exchange and correlation term (Becke 1988; Perdew 1986) has been shown to improve binding energies, yet there is no systematic way to improve density functional theory beyond this approximation. Nevertheless, density functional theory lends itself to computational implementation that could scale linearly in the number of atoms. Furthermore, the simplicity of the basic density functional equations offers great freedom in the choice of the most efficient computational implementation. In contrast, for over four decades *ab initio* quantum chemistry has been dominated by the use of gaussian-type basis functions. The major reason is the necessity to evaluate four-index two-electron integrals in Hartree–Fock theory. This requirement does not exist in density functional theory and thus it can be anticipated that a number of different numerical approaches will continue to emerge, making best use of new hardware architectures. Promising developments in this direction include pure plane-wave based methods with appropriate pseudopotentials (see, for example, Teter *et al.* 1989), non-homogeneous plane wave basis sets (F. Gigy, personal communication), and multigrid numerical approaches (J. Bernholc, personal communication).

An exciting development is the combination of quantum mechanics and molecular dynamics as demonstrated by Car & Parrinello (1985). This type of calculation has become possible through a unique combination of an elegant theoretical and computational approach and the speed of hardware. This approach sets the stage for first-principles simulation of phenomena such as adsorption, diffusion, as well as chemical reactions on surfaces, interfaces, and the bulk materials as found in epitaxial growth, chemical vapour deposition, catalysis, electrochemistry, and corrosion. An important aspect in this work is the unification of hitherto different theoretical disciplines, namely electronic structure theory and molecular dynamics. It can be anticipated that in the future this unification of different sub-disciplines will continue thereby leading to sophisticated integrated approaches.

It is reasonable to assume that in such integrated approaches, different methods and algorithms will have to be applied, each requiring different characteristics of the hardware. Thus, integrated heterogeneous hardware architecture may well turn out to be the most appropriate platform for such approaches. Specifically, there will be massively parallel components, shared memory vector components, and fast scalar processors, together with sophisticated 3D visualization capabilities and access to large external storage devices, possibly using optical storage technology. It is quite conceivable that such heterogeneous compute environments do not have to be physically at one location, but could be spread out even over different continents. From this perspective, the band widths of computer networks becomes an equally important hardware component together with the computer architecture, processor speed and memory size.

Virtual reality technologies as applied to scientific problems are creating great excitement, especially in the non-scientific community. While this is indeed a fascinating development, one should not get carried away by impressive pictures which may or may not represent reality. In fact, one could argue that 'one good number is worth more than a thousand pictures' (D. A. Dixon, personal communication). Despite the impressive progress made over the past decades in the theoretical description and computational predictions of phenomena in solids and on their surfaces, we should be aware of the great complexity and subtle intricacies of even relatively simple systems such as pure silicon or carbon. We have to realize that our tools are still very crude, capturing only small bits of real systems. It can be hoped that as we improve our theoretical methods and the hardware tools, we also increase our respect for the beautiful architecture and marvellous properties of condensed matter.

The author thanks his former colleagues at Cray Research for many fruitful and stimulating discussions, and especially Charles Grassl for the information on memory component developments. Furthermore, the communications with Stephan Brode (BASF) and Michael Wrinn (BIOSYM) have been extremely useful and are gratefully acknowledged.

References

Almasi, G. S. & Gottlieb, A. 1989 *Highly parallel computing*. Redwood City, California: The Benjamin/Cummings Publishing Company.

Andersen, O. K. 1975 *Phys. Rev.* B **12**, 3060.

Andzelm, J. & Wimmer, E. 1992 *J. chem. Phys.* **96**, 1280.

Becke, A. D. 1988 *Phys. Rev.* A **38**, 3098.

Bell, G. 1992 *Science, Wash.* **256**, 64.

Brommer, K., Needles, M., Larson, B. E. & Joannopoulos, J. D. 1992 *Phys. Rev. Lett.* **68**, 1355.

Car, R. & Parrinello, M. 1985 *Phys. Rev. Lett.* **55**, 2471.

Cohen, M. L. & Chelikowsky, J. R. 1989 *Electronic structure and optical properties of semiconductors*, 2nd edn. Berlin, New York: Springer-Verlag.

Delley, B. 1990 *J. chem. Phys.* **92**, 508.

Freeman, A. J., Krakauer, H., Ohnishi, S., Wang, D. S., Weinert, M. & Wimmer, E. 1982 *J. Phys., Paris* **43**, C7, 167–176.

Fu, C. L., Ohnishi, S., Wimmer, E. & Freeman, A. J. 1984 *Phys. Rev. Lett.* **53**, 675.

Hockney, R. W. & Jesshope, C. M. 1981 *Parallel Computers*. Adam Hilger.

Joannopoulos, J. D. 1992 *Science, Wash.* **256**, 44.

Koelling, D. D. & Arbman, G. O. 1975 *J. Phys.* F **5**, 2041.

Kohn, W. & Rostoker, N. 1954 *Phys. Rev.* **94**, 1111.

Korringa, J. 1947 *Physica* **13**, 392.

Krakauer, H., Posternak, M. & Freeman, A. J. 1978 *Phys. Rev. Lett.* **41**, 1072.

Mertz, J. E. & Andzelm, J. 1991 *Cray Channels, Fall 1991*, Cray Research, Eagan, p. 10.

Moruzzi, V. L., Janak, J. F. & Williams, A. R. 1978 *Calculated electronic properties of metals*. New York: Pergamon.

Neckel, A., Rastl, P., Eibler, R., Weinberger, P. & Schwarz, K. 1976 *J. Phys.* C **9**, 579.

Perdew, J. P. 1986 *Phys. Rev.* B **33**, 8822.

Ralston, A. & Meek, C. L. (ed.) 1976 *Encyclopedia of computer science*, p. 1456. New York: Petrocelli/Charter.

Slater, J. C. 1937 *Phys. Rev.* **51**, 846.

Slater, J. C. 1951 *Phys. Rev.* **81**, 385.

Stich, I., Payne, M. C., Kingsmith, R. D., Liu, J.-S. & Clarke, L. J. 1992 *Phys. Rev. Lett.* **68**, 1351.

Teter, M. P., Payne, M. C. & Allan, D. C. 1989 *Phys. Rev.* B **40**, 12255.

Weinert, M., Wimmer, E. & Freeman, A. J. 1982 *Phys. Rev.* B **26**, 4571.

Wimmer, E., Krakauer, H., Weinert, M. & Freeman, A. J. 1981 *Phys. Rev.* B **24**, 864.

Wimmer, E., Freeman, A. J., Weinert, M., Krakauer, H., Hiskes, J. R. & Karo, A. M. 1982 *Phys. Rev. Lett.* **48**, 1128.

Discussion

A. M. Stoneham (*Harwell Laboratory, Didcot, U.K.*): In your excellent user-friendly, general-purpose code you still had a theorist running it for an experimenter. Is the theorist needed in this role? Surely the experimenter wants to do it all.

E. Wimmer: Eventually, experimenters should be able to carry out more and more computations by themselves. However, theorists will always be needed to handle difficult problems beyond routine calculations. In fact, user-friendly computer codes and interfaces will free the theorists to focus on challenging problems thus advancing the capabilities and usefulness of computer simulations while non-experts can accomplish many routine calculations on their own. Clearly, the boundary between advanced and routine work will shift with time.

Index